DESIGNING

THE REAL WORLD

Lon Barfield

BOSKO BOOKS Bosko Books, Bristol, England, UK.

Bosko Books

Bosko Books publishes books and digital resources relating to design for use, new media and information technology. *www.boskobooks.com*

This book is printed on paper from sustainable forests. Bosko Books is committed to lowering the impact of the publishing process on the environment. *www.boskobooks.com/eco*

This edition published 2004 by Bosko Books.
Copyright 2004 Bosko Books,
PO Box 1173, Bristol. BS39 4WZ

ISBN: 0-9547239-1-0

A CIP catalogue record for this book is available from the British Library.

Typeset in Minion 10pt.
Cover illustration, design, typesetting and photography by Peter Maxwell.

To my parents, for showing me the stars
and sea-shells when I was a child.

CONTENTS

PREFACE

PEOPLE AND THINGS

This is a book about people and the world around them; their interaction with everyday things and the spaces in which they live and work. The everyday things are just that; pens, furniture, elevators, light fittings, videos, refrigerators, cookers and more high-tech everyday things including the ubiquitous computer.

This book is also about the design of these everyday things; design for use. At the heart of design for use is the understanding of the relationship between people and things. That understanding acts as an inspiration and a guide. As with many other design disciplines, doing design is not about turning a handle and cranking out formulaic solutions to design problems. Design is about an appreciation of the design problem and a deep understanding of the tools and techniques available to formulate a solution. One vital ingredient for that process is a comprehensive knowledge of 'prior art'; of what has already been done in that area, how well it has worked and how its good points can be assimilated into the design you are currently occupied with. 'Prior art' in interaction design is any interaction that can provide you with abstract information about how interactions work or don't work.

Finally, the advent of interactive, digital media has given us a whole new area of design for use, and an area where the design of the user interaction is more important than ever before. Although this new area of design is based upon new technologies, there is still much to be gained from looking at how people interact with everyday things in the world around them and applying it to this new area of design.

TARGET AUDIENCE

Maybe you are an industrial designer who realizes that there is a lot more to designing a product than just the form and function, or you are a town planner looking to discover ideas and approaches to the subject from neighboring areas. You could even be one of the new breed of digital designers, putting together interactive sources of information on-line. Whatever your background, this collection of columns will have something of interest.

However, the lessons in this book are not just applicable to those who are doing the design, they are also important to those who manage design or those who do the actual construction work; the programmers working on large projects and wondering why there is all this fuss about good usability, the web designer wanting to create sites that work as well as looking cool, the project manager asking themselves if part of the budget should go towards designing the end-user aspects of the system. Anybody whose activities have some connection with design for the user will gain a deeper understanding of what the area is about through the reading of this book.

INFORMATION AND ENTERTAINMENT

There is an assumption among some people that if something is to be learned it should not be amusing and enjoyable, rather it should be a chore. This is both unsavory and untrue. Indeed, the converse is true; we are more likely to remember something if it is embodied in a story or anecdote and especially so if it is amusing, sexual or bizarre. Whilst a Kama Sutra of interaction design is a challenging proposition I have limited myself to including a fair degree of amusing and bizarre anecdotes in this collection in the hope that they will help the readers get to grips with the ideas and, more importantly, remember and apply the lessons they contain.

In conclusion, this is not meant to be a funny book, rather it is a book that is intended to be 'amusing yet authoritative'. It is an attempt at that eternal goal of educationalists; teaching people things without them being aware that they are being taught. So; read on and enjoy the columns and let the learning look after itself.

THE REAL WORLD COLUMN AND THE SIGCHI BULLETIN

Some of this book is based upon the 'Real World Column', a regular column that deals with interaction design in the world around us. It was started in 1995 in the SIGCHI Bulletin, and is still running today in online form. This book brings the columns together in one place, organizes and illustrates them and supplements their content.

The discipline of Human Computer Interaction (HCI) emerged in a big way in the late 70s and early 80s. The Association for Computing Machinery (ACM) in America has many special interest groups (SIGs) active in different areas of computing. The one that was born as a result of the HCI movement was SIGCHI (Special Interest Group in Computer Human Interaction) and, like the others, it had its bulletin; a quarterly newsletter of articles and news. The rise in interest in the area led to the bulletin going bi-monthly and eventually the creation of a new magazine 'Interactions' which included commercial interests and also placed more emphasis on interaction with things that weren't computers.

Early in 2003 the print version of the bulletin was brought to an end, its role having been superseded by 'Interactions' magazine. At that point the 'Real World' column moved to its own dedicated web site (*www.idhub.com/realworld*) and the decision was taken to publish the content in this book you are now reading.

The book has much more than the web site. Whereas the web site is simply a collection of the columns, the book brings them all together in a coherent whole and adds accompanying text to bind them. There is also much new text in this book in the from of previously unpublished columns and a large introductory section that deals with interaction in the real world and its observation. Furthermore, this is a real book, you can hear the pages flick and smell the ink and paper, none of that insubstantial, digital stuff!

THE BOOK'S STRUCTURE

Incorporating a collection of columns into a book is a classic example of repurposing content; taking content that was initially written for one context and placing it in another context. The columns published to date have been reordered within this book to give a structure that is more logical. Readers wanting to see the original order in which they were published should consult appendix C. The structure of the book does not reflect an ideal structure for the subject of real-world lessons for interaction design, rather it is an optimal structure for the content available in the columns. The structure of the chapters is as follows:

INTRODUCTION: The book starts with a look at interaction in a general sense and how much of our lives is made up of interaction. It then looks at how interaction can be designed when it is embodied in the behavior of technical products and systems.

THE REAL WORLD: The real world is considered in an abstract way as a collection of different types of interaction. Further consideration is given to three key types.

LOOKING TO THE REAL WORLD: The introductory section ends with a look at how we can extract knowledge of interaction design from the real world around us, both the ways of approaching the task and the types of things we are looking for.

THE COLUMNS: This is the main part of the book. All the already published columns and a number of new columns organized into the following themes:

HARDWARE: This covers the low-level, building blocks of hardware interaction; buttons, switches and the like. It also looks at managing powerful functions in devices and at assorted examples of hardware interfaces.

PEOPLE: In 'People' we look at aspects of the relationship between things in the world and people using those things. We also concentrate on some aspects of being a 'person' that are important when doing interaction design.

TIME AND NARRATIVE: 'Time' deals with aspects of interaction that are tied into temporal effects and 'narrative' concentrates on abstract aspects of narrative that could be used in the interaction design arena.

INTERACTION AND SPECIFICATIONS: Abstract topics of interactions are covered in this section such as terminology and loops, while the section on 'specifications' takes a quick look at the formal side of describing interaction, including e-commerce and the applicability of forms to business scenarios.

CONCLUSIONS: Finally there is a summing up of what we have learned – not so much the individual lessons but the overall thrust of observing the real world as interaction designers.

THE ACCOMPANYING WEB SITE

Many of you will have arrived at this book through the web site, but for those of you who have not, there is a web site at *www.idhub.com/realworld* where new columns are still being published on a bi-monthly basis. If you enjoy this book then you can visit the site to see the latest additions or add your email address to the mailing list and have the latest columns sent to your inbox as soon as they come out. This web site is actually part of a larger collection of interaction design resources, some of which are described at the end of this book.

ACKNOWLEDGMENTS

Firstly, thanks to Steven Pemberton, past editor of the SIGCHI Bulletin for allowing me to start the column. Then, thanks to the various SIGCHI Bulletin editors and assistants who kept the whole thing ticking over. For this volume thanks go to Alix and Cheryl for initial proofing, Greg for feedback and to Pete Maxwell and everybody else at Bosko Books. For the general context I am indebted to Morgan and Keiran for supplying the kids view on technology, and finally Wendelynne who, for years, has put up with me moaning about usability in stations, airports and cafes in Europe and America.

INTRODUCTION

LIFE IS INTERACTION

A researcher leans over a huge pool with two rubber pads to mask the eyes of a dolphin for an experiment on sonar. One of the rubber pads, slippery with water, falls from her grasp into the pool. Like a knife, the dolphin dives after it and returns to the surface with the rubber pad in its beak, held clear of the water. The dolphin bobs toward the woman and tilts its beak to offer the pad. She reaches out to take it, but as she does so the dolphin bobs away from her slightly, just enough to keep the pad out of her reach. This is not coincidence, because the dolphin bobs nearer again and the process is repeated. This is play, this is teasing, this is interaction.

Two removal men are negotiating a bend in the stairs while carrying opposite ends of a small but heavy table. In the background a radio is playing loud pop music. As they edge it round the balustrade they pull and push at the table. Partly this is just moving it to avoid bumps and edges, but partly they are telling each other about what they are doing and what they want to do. Unable to gesture with their hands, not wanting to waste time with complex verbal explanations, they are communicating their intentions regarding direction and orientation by shifting the table and feeling the way that the other is trying to shift it. This is interaction.

An old man goes into a small booth, looks nervously around him and uses a large machine to punch a hole in a list of names. It is the first time in generations that voting has been allowed in his country. There is an interaction with the chunk of equipment in front of him, but it is miniscule compared to the interaction that takes place between him, others like him, and the leadership of the country. The metal card punch is a system but so too is the abstract structure of the electoral system. This is interaction.

1

In a brightly-colored market a thin woman, with a blue hat and a red sunburned neck, holds up four fingers to the stall holder. His teeth flash white, like a row of Lego bricks, and he shouts something to the small boy helping with the stall. A scarf in vivid reds and purples is added to the three that are spread out in front of the woman, she smiles and the blue hat tilts as she nods in approval. This is interaction.

A conductor is sweating under the lights and his heavy black jacket. Off to one corner of the auditorium the organist sits in a dark, wooden booth beneath a tumble of giant tubes poised over her like a sword of Damocles. In a world of her own, her two links to the outside are the sound of the orchestra accompanying her and a small black-and-white screen mounted above the stacked keyboards of the organ. On this screen the conductor swings his body and slices the air with his arms as the music reaches a crescendo. This is interaction.

A man with a gray beard grins and slowly moves his hand. He picks up the black piece and moves it to the edge of the board. His younger opponent gazes with the intensity and narrowed eyes of a polar explorer scanning the horizon. Gradually a ripple of applause and exclamations trickles through the small crowd as they read the strategy unfolding from this move. This is interaction.

Is there anything that takes place in this world that is not interaction? Is there anything that happens, that ever happens, in total isolation from everything else? Isn't this world a network of interactions; a tapestry of meetings and exchanges of information within which we are just one thread?

Interaction is about being somewhere and doing something. It is as broad and as rich as life itself and it is only by studying this rich world of real-life interactions that it is possible to make informed decisions when designing the interaction embodied in an artificial system. Interaction in this wide context is not just the interaction of man and machine. It is more than buttons, levers and display screens. Interaction in this context is interaction between anything and everything.

INTERACTIVE LIVES

Above we saw that our entire lives are made up of interaction of one sort or another. In the vast majority of cases this interaction runs smoothly and effortlessly, sometimes so smoothly that we are barely aware of what is happening. That typifies most of our interactions in the real world, the ease and fluidity of the interaction. The situation changes when we look at those aspects of our interactive life that are artificial and designed by people.

Most of us at some time in our lives have lost our temper with a piece of everyday technology. It might have been the photocopier that tries to copy only half of our document, the elevator that is going down when we want to go up, the automated phone service that won't tell us the one thing that we want to know, the alarm-clock that decides not to wake us up in the morning. More often than not it will have been the video recorder that recorded a baseball match instead of the second half of a Puccini opera.

Sometimes our anger won't be directed at the technology, but at ourselves for not doing the 'right thing' with the technology; coming back from the holiday to find the heating still going at full blast, not backing up our laptop until it goes missing, putting the wrong chemicals in the dishwasher, or recording something over the first half of the Puccini opera because we forgot to label the video cassette properly.

The complexity of life

Our lives are awash with systems and technologies: at work, at home, in the car, in the street, even on our person.

The way we live has dramatically increased in complexity in the last few decades; I know my grandmother didn't really have the hang of the phone or the remote control, and why should she? They were not part of the life she grew up with, they were a part of what that life had become, but for us this sea of complexity that we swim through has become second nature. From the moment we reset the alarm-clock on waking, to the moment we switch off the heating when we go to sleep, we are continually searching, programming, setting, activating, editing, reviewing. We have grown up with it and we live with it so closely that we forget how complex it all is. That is until we have to explain it to someone else. If you have ever been on holiday and have had someone house-sitting or cat-sitting then you become only too aware of the complexities and idiosyncrasies of your living space; where the heating control is, how to operate it, how to operate the timer for the lamps, accessing the internet, how to heat the water for a bath, how to operate the alarm system. And that is all for someone who is familiar with the underlying technology. Imagine trying to explain it all to someone who wasn't.

There is a strange paradox when it comes to complexity and detail. In interactive systems too much complexity is usually a bad thing, as the user has to get to grips with it. This is true of both the internal complexity and the external complexity: complexity in the presentation of information. Now, the interesting thing is that complexity and detail are not inherently bad things. Nature is infinitely more complex than any aspect of our cultural invention. The simplest tree is such a complex system both in the aspects of it that we can't see and in the parts that we can see. To have a hard day behind the computer struggling with obscure programming languages and then to step into the countryside is like a breath of fresh air (quite literally) but why is the complexity of nature such a relief after the lesser complication of the world of computers and things? The answer must lie somewhere in our history as natural beings. We have evolved in environments that are natural, our species was born and lived and matured surrounded by these things, so just as a fish probably breathes a sigh of relief when put back in the water, we too breath a sigh of relief when we step back into the world of nature.

The question then becomes how do we remove the artificial complexities from the systems we design, or how can we make that complexity more 'natural' in its qualities. In the world of housing design where houses are getting continually smaller and simpler in layout, there was a report recently that criticized this development. It compared older houses to today's modern dwellings and interviewed those who lived in them. The results were that houses with more nooks and crannies, more little quirks and features were more favorable to the inhabitants. The space was more flexible and felt bigger because of the detail inherent in the house. There were more gaps for cupboards and small spaces under stairs and the like. The report concluded that modern houses should be less functional and efficient and should include more of these little details.

So humans need some forms of complexity while other forms are problematic, especially in the world of interactive technology. How then do we cope with this complexity in interactive technology? Our ability to cope is helped by three main factors:

Slow change

Firstly, these things have changed slowly. The slower the pace of change the better we can cope with it and although it seems like the technical product landscape has been turned on its head in the last twenty years this has happened at a certain pace.

Companies can lead in that pace, they can be the first ones to bring a new product to market. They can also dictate that pace to a limited extent, they can raise the profile and thus the demand for a new technology by skillful use of advertising and marketing. If the company gets it wrong, if they try and introduce a product faster than the pace of change that people can cope with, then that product will flop. Something akin to Apple's Newton, it was first on the block with a PDA but the world wasn't quite ready for it so that was one of the factors that caused it to flop.

Better coping

Secondly, the two points above allude to things that make it easier for us to cope with increasing complexity and that in itself is also a key point; we can cope. Furthermore, as the complexity increases, so too does our experience of complexity and thus our ability to cope with new complexity; the strange little icons on the back of our new digital camera are just like the icons we are used to on our desktop, programming the heating is similar to programing the video recorder and so on. Just as designers are creating new systems that are adapting to the needs of the user so too are users adapting to the needs of the technology.

Better design

Thirdly, the whole area of designing, managing and presenting complexity to the consumer has gone through a shake up. As technology has become more complex, companies have realized that usability is a key marketing tool and so they have invested in research and development in this area. Yes, things are getting more complex, but that increase in complexity is being matched by an increase in the design effort to tame that complexity and make it more palatable.

INTERACTION: ONE ISSUE FROM MANY

Finally, it should be stressed that the meeting point between people and things is a very wide one and includes many issues; the relationship between users and the things around them, users and the rise of the consumer society, the control of technological research and development by large corporations, the marketing and selling of products, the whole ownership thing, how we define ourselves by what we have. It is a rich area indeed, bringing together the most salient aspects of technology, people, society and personal politics. Some of these issues will be dealt with in another book currently in production dealing with the 'big picture' of people and things. However, this text will concentrate down onto the interaction between people and things. There will be no big value judgments beyond saying what is good interaction and what is bad interaction.

THE DESIGN OF INTERACTION

Above, we considered how much of the world around us is composed of interactions. Interactions that work efficiently and smoothly. When we compare this to the designed world; the world of technology and computers, we realize that these interactions are far less satisfactory. Comparing these two observations begs the question; 'Why isn't it possible to design the second world so that the interactions are as good as in the first world?' Well answering that question is what interaction design is all about.

What is interaction design?

Interaction design is design, it is the design of those parts of a system that the user has some contact with. It has long existed as an unrecognized part of many disciplines that were connected with design for use. Industrial designers were key players with their issues regarding the Man Machine Interface (MMI). Town planners have also had to deal with the problems of how people on the street perceive and use their designs. Graphic designers too have used it in various forms. A graphic design can communicate on the levels of emotion, style and beauty, but if it cannot communicate on the functional level, if it has difficulty actually communicating the informational content, then the graphic design fails.

The subject of interaction design only really came to light when we humans shifted our attention to the ultimate in complex, interactive products; the computer. A fledgling discipline was born called 'HCI'; human computer interaction, or sometimes 'CHI' computer human interaction (same words but the acronym is easier to say – a key factor if the discipline is meant to concern itself with how end-users get on with the things that are being designed – even an acronym has to be usable). The fluid nature of the area led to many different terms coming and going, but mostly coming since there are still many different terms in use and the discipline itself is expanding and changing.

I shall stick with 'interaction design' for this book, as the other terms are more specialized to the field of people and computers. What I want to deal with is the interaction between people and – well – anything else.

Interaction design is a specialist discipline aimed at designing all the aspects of the dialogue between the user and the system, and that means all. If we consider an interactive product then that dialogue with the user begins before they have even got the thing, they see others using it or they see it in the press or on the web. The dialogue continues to include the publicity material and the box. Then the packaging, the manual and only then does the interaction design of the thing itself start to play a part. Further down the line we also have interaction design playing a part in the maintenance of the product and in the errors, troubleshooting and help-line when things go wrong.

Can interaction really be designed?

The simple answer to this is: 'yes'. Although interaction seems like a very tenuous and vague thing, it is influenced by many concrete factors: the language used at the interface, the visual design of the interface, the way things change as the user steps through from one stage to the next.

Although interaction design is 'made visible' by all the tangible factors that it depends on there is always the danger that the design of the interaction becomes eclipsed by these more tangible aspects. Designers may become preoccupied with visual design for example, not as an ingredient of interaction design, but purely as visual design. You then end up with a situation where the designers are pouring time into making it look good, the client or manager of the project thinks that it looks good and the end user will say, 'Wow! That looks really good.' Everybody is happy with the project until someone comes to actually use it and realizes that using a system is very different to just sitting in front of a screen going, 'Wow, that looks nice!'

Interaction isn't only intangible when compared to aspects such as visual design. Interaction design also needs more time and effort to be appreciated. The evaluator needs time to interact and build up a feel for the interaction, sometimes they need to carry out the interaction several times to get a good idea about what is going on. In such circumstances it is easy to see why students, lecturers, developers, clients and even designers can become side-tracked by the more tangible aspects at the expense of good interaction design. It is a struggle to concentrate on the interaction aspects of the design and it is a struggle to make others concentrate on them as well. Execute the interaction design well and the system will be a pleasure to use. More than that, the users will hardly be aware that they are using it as it will flow and react so naturally. Design it badly however, and the system can be rendered unusable; either people will avoid all contact with it or, if they have to use it for some reason, they will do so grudgingly and all the time curse whoever it was who was responsible for the design.

Is it just applicable to new technology?
People who talk about 'new media' often talk about 'interactive media' or 'interactive, digital media.' It sometimes seems as if interactivity has only just been discovered. To some extent this is true; with the advent of complex digital systems we have certainly become more aware of interactivity. However, this does not mean that systems before that did not have interactivity. Every system that the user has some control over can be said to be interactive. Before interactive digital media there were plenty of non-digital, interactive systems. Indeed there still are, just think of the video recorder, the heating controls, the car controls etc. Even simple things like doors, faucets (water taps), lights and pens have a degree of interaction with a user and there are always examples of both good and bad design.

THE REAL WORLD

NATURAL, DESIGNED OR DIGITAL

The world around us is a wonder. The depth of complexity, detail and beauty is incredible. This detail is a source for all manner of ideas and inspirations, interaction is just one of them. When we consider interaction further we will need some classifications to help us. Let us start by dividing it up in terms of the three worlds: nature, designed and digital, and considering each of these worlds in terms of the things that inhabit it and the agents that operate within it.

Natural world: physical things

Firstly, we interact with the natural world around us and we have been doing so for a long, long time, as the world around us has gradually evolved so too have we evolved to be able to perceive and react to certain things that happen around us.

Thus, what we see today is highly influenced by how our eyes and visual systems work and that, in turn, is dependent on how they have been shaped by evolution to notice things around us in the natural world. We are sensitive to the color of berries against leaves, since this ensures that we can find and eat such food easily. The result is that our eyes are highly attuned to red, and red is perceived as an attention getting color, easily noticeable against the background. Use is made of this in design, and red is used as the color of warning. Not only that but objects have been found to be more noticeable when displayed against a green background so the combination of red against green is the most effective for noticeability – definite parallels with berries on undergrowth in nature.

But it is far more than just the associations of color, all the other senses are finely attuned to finding food or avoiding being food for something else.

Natural world: agents

There is more to the natural world than just rocks, plants and berries. We also interact with animals and, to a far greater extent, with other humans. Man is essentially a social animal. As well as evolving opposable thumbs and standing upright, we have also evolved societies and a common understanding of the world around us. These are just as important to our survival as our legs or brains.

Communication with other humans takes a multitude of forms and it is also interesting to note that many of today's technological advances are connected with communication. This may be in the form of face-to-face meetings or remotely, and either one-to-one meetings or one-to-many.

Designed world: physical things

We now move onto the designed world, but in particular the designed physical world. The world of products and gadgets, of buildings and streets, of posters and books. The design of these things owes a great deal to how we perceive and understand things around us and thus to how we have been shaped by our natural evolution. Take the use of yellow and black for communicating a warning, it is no coincidence that this is the same color combination used in wasps and bees. Although the designed world is still far away from the natural world there are still some shared elements, the designed world is still a physical one, we can still see and feel the elements, we still push and prod the controls in the same way that we used to push and prod stones and animals. A good understanding of how people interact in the natural world is a vital grounding for the design of interaction in the real world.

Designed world: agents

There is a bit of a gap here; what are the designed equivalent of agents? We certainly don't design animals and other humans to interact with, but gradually we are building up understanding about artificial agents that are independent from humans and with whom we can interact. They range from the simple, robotic vacuum cleaners through to semi-intelligent voice-operated booking systems.

In both the designed world and the digital world it is easy to produce 'inanimate' things that the user can shove around, whether it be a dumb cardboard folder or a dumb icon of a folder on the screen. The idea of 'agents' in both these worlds is a different matter and the dividing line between the two worlds is a lot thinner. Any agent-like qualities are by definition going to involve digital processing. You can't make a semi-intelligent vacuum cleaner without including a little digital brain. So this category of 'designed world: agents' is not really a valid one and the real discussion is about agents in the digital world coming up later.

Digital world: physical things

Finally, we come to the third area of design, the design of the interactive and digital world. 'New media design' as it is sometimes called. This is a part of the designed world but it is more remote from the natural world because its elements are more intangible. Things can still be seen and heard but the channels of communication are more tenuous and every aspect of them has to be designed. The design of this area draws on how we interact in the natural world but a large proportion also comes from a study of how users interact with things in the designed, physical world since this is already a part of our historical interaction context. Although it is a very small fraction of our interaction history it is nonetheless the most recent part and thus has an importance out of proportion to its size.

Digital world: agents

This has already been discussed in the preceding sections. The whole area is still very much in its infancy and our first steps towards intelligent agents have been controversial; the Microsoft Paperclip is a real 'love him or hate him' figure, and it will be a long time before we interact with the likes of the HAL 9000 computer in 2001: A Space Odyssey.

Having said that, there are systems that are already shifting from the research world to the commercial world. It is claimed that some pornographic chat services (where you pay to flirt in text form with young men or women) cut costs by charging you to flirt with a digital agent that mimics flirty chat.

Three tangible worlds

I have described above the six worlds that can be identified by an analysis of interaction. Of these six worlds, the last three are more erudite since they are concerned with digital technology and with agents. The remaining three, the first three in the series, are of more interest to us in terms of studying the real-world because they are more tangible. They correspond closely to key areas of study when considering interaction in the real world.

The first category, that of; 'Natural world: physical things' corresponds to the world of nature. How we interact with well-defined, natural, physical things in our environment. The second category, 'Natural world: agents', corresponds to the world intelligent agents in the environment around us and that is of course people. The final one is 'Designed world: things.' This corresponds to the world of artificial things in our environment. These three areas deserve more detailed coverage.

THE WORLD OF NATURE

One could argue that we don't have much to do with nature in today's modern high-tech world. Even when we do occasionally venture out into real nature we are swathed in water-proof jackets, well-designed boots, high-tech fleeces and usually carry a few bits of mobile technology along with us just in case: phone, watch, camera etc.

However, this distancing from nature is a very recent thing in evolutionary terms. As far as our genetic make up is concerned we are still very much designed for hunting and foraging in the wilds of nature.

When I first heard Darwin's evolutionary theory summed up in the phrase 'survival of the fittest' I took it to mean survival of the strongest, the healthiest. In fact what it means is survival of the one most fitted to the context. Strength does play a role but so too do lots of other factors. If the trees that the creature feeds on are all of an average height then having a neck that is too long will be as detrimental as having one that is too short, the majority of the leaves will be out of reach. The creatures most suited to survival will be those with the length of neck that corresponds to the average height of the tree, the ones most fitted to the environment.

Humans are no different. It is true that recently (perhaps the last couple of million years) they have found ways of adapting their environment to suit themselves. However there are many ways in which we are still the product of the millions of years of evolution that preceded that. Evolution that made us fit into a certain context. Our physical and behavioral programming is thus fine-tuned through thousands of generations to match a definite environment and set of tasks. This is also true of our physiology in terms of our senses. Although we have learned to process visual and audio data in complex ways in the last few thousand years, the fundamentals of the way that we see and hear have been laid down over millions of years of evolution. How we coordinate our vision and actions, even how we search and find things and navigate through our environment is all hard-wired in. This legacy of physiology and behavior is a great pool of resources that multimedia designers can take advantage of.

So, on a basic level we are 'designed' by evolution to fit with our recent evolutionary context, that is our context before the development of our ability to think abstractly. Now we can think abstractly and even override some of the in built behavioral programming. We can go along to the dentist and wait quietly even though we know they are going to hurt us when they fill that tooth. When the big guard dog barks we will jump but we will not flee because we know that he is on the other side of the fence. The most amazing part of our separation with evolution is that we now have the ability to alter our environment in pre-meditated ways; we design. It therefore makes sense when designing our context to design it to fit in with what humans are used to in an evolutionary sense, to incorporate certain factors from the natural world that match with what we ourselves are designed for. In effect we should mimick the natural world in our artificial designs because the natural world is what we were made to process and understand.

THE WORLD OF PEOPLE

Interacting with people is something we learn about right from an early age, and to some extent even earlier, since some of our interactions are based on deep-seated genetic factors. We interact with the people around us for a variety of reasons. Consider the following:

COMMUNICATING: This is the true exchange of information. It includes things such as giving instructions and explaining things. I suppose what I am trying to distinguish here is the one-sidedness of the communication. It is not about communication to reach a consensus or develop ideas, it is about communication as a means of transferring information. There is an idea in my head and I want to put it in your head as quickly and as efficiently as possible.

DISCUSSING: Discussing is, to some extent, a part of communicating, but it is less direct and more open ended. Discussion is about the exchange of opinions on some subject and the way those opinions influence others. It is about exploring aspects of an issue and it is about learning.

MAKING IDLE CHIT-CHAT: Very often, interaction just plays the part of 'social glue'; communicating with each other for the sake of communicating with each other, exchanging pleasantries or discussing the weather. There is no direct purpose served in terms of exchange of information, but the fact that there are two people communicating with each other is important in itself.

STRUCTURED DISCUSSIONS AND MEETINGS: Finally, there is the world of structured interaction, interaction that is pre-arranged to achieve a certain goal. One can think of the many forms of interaction that occur between people in the business world: the interviewing of candidates for a job, the meeting which follows a formal agenda of points, or the business presentation with questions afterwards.

DEEPER FUNCTIONS OF COMMUNICATION: One of the points above was 'idle chit-chat' – conversation that served no direct functional role. The phrase 'idle' is misleading as this type of communication is one of the most important that there is. Interaction is not just about getting information from one brain to another. There are deeper things at play in human-human interaction. As we interact, and as others interact with us, we use this interaction to help define our place in the world. The perception of our importance in the world is governed by the web of interactions that we have with those around us and, apart from a little bit of genetic pre-programming, the essence of what we are as an individual is defined and shaped by successive layers of interactions, with each layer affecting how we interpret and are shaped by the next interactions. We are the product of the sum total of all our interactions.

The channels

There is not just a wide range of roles that interaction plays, there is also richness and diversity in the different channels of communication used by humans. When we think of the communication that takes place when two or more people interact we usually think of speech; the things that are said. In fact a huge amount of communication is carried out by means other than speech and although speech is the best medium for communicating complex information accurately, the messages that are communicated via the other channels can be more powerful.

Humans seem to have the ability to utilize every possible channel available for communication. If there is a means by which a human can communicate with another human then they will make use of it, from the simple flashing of headlights in a car through to complex designed systems such as the tic-tac men signaling the betting odds at racecourses or the complex, gay hanky-codes used to signal sexual preferences.

Looking at the more conventional channels of communication and interaction we can divide them up according to our senses. There is much that could be said about interaction by means of scent and touch but in terms of designing systems we shall concentrate on the more 'designable' channels of vision and sound.

When we talk about sound, we mean the human voice, all the nuances of the voice that are layered on top of the actual information content. The timbres and tonalities, the speed (or hesitancy), the accents and the emphasis.

On the visual side of human communication, there are many facets. For starters the eye itself and the issue of eye contact. Eye contact is a vital part of interaction, as the saying has it, 'the eyes are the windows of the soul.' Eye contact and how it is maintained or otherwise is one of the key players in Freud's ideas of unconscious 'nonverbal leakage': The way in which attempts at deception leak out of us like water from a broken bucket.

Moving on from ocular communication, there are facial expressions and gestures. These can be consciously adopted but more often than not they are used almost unconsciously. One only has to watch someone having an impassioned phone conversation to see a whole array of redundant expressions and gesticulations. Facial expressions and gestures are a form of communication that is somehow much deeper than spoken language. Language can change and adapt over time as areas are conquered by new empires and new languages enforced, but gestures are more deep seated and tend to stay in place across the centuries. To feel how ingrained gestures are try saying 'I've got no idea' without shrugging or try saying 'no' while nodding or 'yes' while shaking your head.

Physical contexts

Our interactions with one another all take place within certain contexts and boundaries. The most apparent of these is the physical context, and because the physical context has played a role in our interactions for many millions of years, the effect it can have is very pronounced.

Physical context covers issues such as territory and personal space. These issues may seem appropriate to animals in the wild but are they relevant to us humans? Well, one of the reasons that they don't seem all that relevant is that the issues are very rarely tested, so well managed and organized is our social structure. Our physical space is secure and well defined and the people around us follow the unwritten rules of interaction very closely, so there are few occasions where we have to scare others off our patch or fight over a female of the species. However, on rare occasions where territorial issues do arise it can quickly lead to tempers and arguments.

Personal space is the same, we all respect one another's personal space in our interactions, but go to a different culture, where the unwritten rules are different and you can have a completely different experience: in France the personal space is smaller; people get closer to one another and they find the English and Americans distant and aloof while we can find them too much 'in your face' physically. In Poland the rules for passing people in the street are subtly different; if you go there on holiday you will find yourself bumping shoulders with strangers, or bobbing from side-to-side as you both try and pass each other.

Social contexts

As well as the physical contexts in which interactions take place, there are distinct social contexts as well. Consider the following three main categories of interaction, and think about the rules and styles that the interactions take in the different arenas:

HOME: Our home is a very special and private place, we interact with direct family members, our partner and our children. We also interact with temporary inhabitants; extended family members, friends, visitors. Beyond that, there are the people in the vicinity of where we live; the neighbors and the people that provide services like the milkman or the mailman.

WORK: In our place of work there are also a large group of people that we interact with. Indeed a huge proportion of our lives is spent interacting with them. There are similar shells of closeness to those identified above; close colleagues, people working above and below us in authority, clients and external staff.

SOCIAL: The third big arena is the social world. Getting out and meeting up with friends and acquaintances, doing new things and discussing all manner of subjects. This area of interaction is very different to the interaction at work; we are able to be more relaxed and take more risks and the context is more supportive of our views.

As an aside there are also extremely casual yet regular interactions with unknown people in our lives. Interactions that may be minute in terms of exchange of information but that assume an importance because of their regularity; there is the old man you always nod to on the train, or the young lady in the van that you exchange smiles with each morning as your vehicles squeeze past each other in the narrow country lane.

There are four main sources of interaction style

Where does our skill at interacting with others come from then? There are four main factors that govern it:

The first, and probably the most important, is what governs the deep-seated, in-built factors: the smiles, the unconscious body language and jostling for territory. The ground rules for these have been laid down over millions of years of evolution, some of them we are completely unaware of, others we are aware of but powerless to change.

Although these factors are deep-seated, humans are in the unique position of being able to think and reason about them and to change or influence them to some extent. This conscious use and manipulation of these forms of interaction enables us to overcome some aspects of our 'programming' and even to take advantage of the programming in others in order to achieve goals within an interaction.

A third source of interaction styles and patterns is the one that we learn from the culture we live in. Within the different cultures on the earth there are many different ways of doing things, and – just as interestingly – many things that are done in exactly the same way.

Finally, there is the more formal version of the cultural behavior described above, the structured ways of behaving for certain well-defined occasions. These are the interaction forms that are studied and adopted almost in the same way as one dons clothing for a formal occasion.

Human-human interaction as a design tool

Using interaction between people as a metaphoric scenario is useful when showing problems in designed interactions. If you are struggling to explain to a management team how unacceptable a certain interaction design is, you can simply say, 'Imagine if you got the same treatment from a person.' We are so used to accepting second-best interactions from technology that it is sometimes difficult to see what the drawbacks of a particular interaction design are. Seeing the same interaction cast in the light of human-human interaction can make it painfully obvious what the problems are. I often used this approach in lectures, and it was the basis for the Hemelsworth and Barker sketches that appeared in my book 'The User Interface: Concepts and Design'. (ISBN 0954723902)

THE WORLD OF THINGS

A friend of mine once made the assertion that the key things in a person's life are religion and politics. To some extent this could be true; politics is concerned with economic equality and justice within the society and religion gives you tools for making sense of life in general. Consider though the state of affairs here in the Western world; religion has become a pale shadow of what religion is in other countries. In surveys, huge numbers of people declare that they follow a particular religion, but precious few actually go to church or practice things connected with that religion. Religion has lost the spark that it once had. It is a similar story with politics, with the current trend towards democratic liberalism there is very little to choose between the main parties; couple this with an electoral system that embodies proportional representation and you have very little of the political passion that once moved the world of politics. Politics too has lost that spark.

What are the important things in life in this Western society then? One of the main ones is quite simple; the things we surround ourselves with. Here in the West we bob along in a veritable sea of 'things'. To some extent this is a bad state of affairs; to consider that the world of tangible things could be more important that the intangibles of religion and politics and everything else. Does this mean that designers should stop now and all go out and become missionaries and impassioned politicians? Not all of them in my opinion, there are many factors that affect our lives but whatever is happening we will still have 'things' and the quality of that part of our lives will be governed by the design of those things.

Not only that, but all these divisions I am describing are never as black and white as they appear. To some extent there is always a bit of politics in design. Designing a product that is mass produced and that comes into close contact with people can be likened to making a political broadcast. With one action you have the chance of reaching millions of people, you have the opportunity to show them things, to influence the way that they think and live their lives, albeit in a minute way.

And there is certainly a bit of religion in design, 'stop and consider the works of God' is a common refrain in a number of religions; stop and look at the wonders of the natural world. Very rarely you get designed systems that have the power to elicit this same awe and wonder; a grand statement of a building; a perfect product; a small feature that shows the designer is thinking of you, the user; an exquisite piece of handiwork from a time when people invested time in the things they made. There were tears when the supersonic plane Concorde was finally laid to rest. People said it inspired them, it was a thing of beauty, revered by many, a vision, a thing of perfection, it had the power to move those that saw it. I am not holding up design and religion and comparing them and saying that design is like religion, what I am talking about is finding aspects of religion in design. Not in a very well defined way; you can't find aspects of Christianity in your toothbrush or Zen Buddhism in the light switch, but you can strive to instill your creations with that same sense of awe and wonder and strive for that feeling, as a designer, that you are doing 'the right thing'.

The best example of this amalgam of design and religion is the craft-work of the Shaker movement. These days people with disposable income are buying reproduction 'Shaker style' kitchens, with plastic legs hidden behind chipboard hidden behind painted veneers, with cheap polished fittings. They are 'Shaker style' only in the superficial appearance of the outer layers. The Shaker movement in the 18th and 19th century embodied many philosophies, one of which was worship through handiwork, not just 'keeping the hands busy' but producing quality items of simple, good design; 'Put your hands to work, and your heart to God.' Their work was governed by a belief that God could see all the details of the manufacture and the finished product. Even if a bad wooden joint was covered by paint, or hidden within the base of a table people probably wouldn't notice it, but the Lord would. The result was a noteworthy contribution to culture in the United States through their architecture, furniture making and handicraft. They produced a range of simple, exquisitely made products and achieved a global name for quality that is the envy of any of today's multinational corporations.

The question I always ask myself is if the same thing is possible with interactivity. Can we envisage such a thing as a 'digital Shaker movement'? Is it possible to design interaction that instills the same feelings as Shaker products do? The act of designing with reverence to a God may be misplaced in today's society but perhaps we can still design with reverence, if not to a God then to the end user. One way is to think about what real-world interactions posses 'quality', analyze them and then reproduce that same effect in our designs. If I think about the most quality real-world interactions I have had with anything at all then the list would look something like this:

Brainstorming with someone of a like mind and both hitting upon the same idea at the same time, that moment when you are both trying to explain it and both saying 'yeah' at the same time.

Dancing 50's style rock-and-roll, you put your partner through a series of spins, turn your back, stick your arm out in time with the music and at exactly the same moment your partner is putting her hand out and 'whack' you grab each others hand.

Playing music together, improvising, weaving in and out with snatches of melody, building up to a sudden stop and then with a bit of eye contact bring the whole thing to a crisp ending in perfect synchronization with each other.

Now, if you could build those same feelings into a designed product then you would have something that people would love to use. Are there any products like this? Is there some bit of technology that when you use it you get that fantastic warm feeling inside? I can think of precious few, but sadly I can think of hundreds of products that elicit exactly the opposite feeling.

L O O K I N G
TO THE REAL WORLD

Finally then, we come to the central thesis of this book: the real world is the richest source of new ideas for interaction design.

Our lives, as members of the human race, are built upon interactions of all sorts. Barely a moment goes by without us taking part in some form of dialogue, be it with an object, an agent or an environment. Interaction design is not just about applying theories while we are at work between nine and six. It is about continually being aware of interaction and continually gleaning ideas and insights from interactions in the world around us. It is about trying to migrate those ideas that work to the systems that we are designing and trying to avoid those ideas that don't work, so that we can create interactive products that add something to peoples lives.

Being an interaction watcher is not something that you can learn overnight, below are a few pointers for the first steps. After them I shall look at how key ideas and issues can be incorporated into interaction design practice.

BUILDING UP KNOWLEDGE OF INTERACTION

I keep referring to observing interaction in the real world. In fact observation is just part of the picture. There are four key ways of amassing knowledge of interaction design: by watching, interacting, talking and designing.

Knowledge from watching

Good observation requires the development of a 'third eye'. A designer's eye that watches whenever your own two eyes are tied up with trying to read information or watching as your hands fumble for some strangely shaped tap. Your designer's eye can watch all this and make judgments about what is happening while your own eyes and hands are cursing whatever dumb designer put the system together. This approach also has the hidden bonus that the more frustrating and stupid an interaction is the more you can say, 'Wow, what a classic example of bad interaction design.' Personally I am sure that this approach has already lengthened my life by several years due to reduced stress levels at such fraught times.

Maintaining this vigilance is easy when it comes to bad interactions. What is trickier is remembering to be observant when you hit a really good interaction. A good interaction is usually typified by being so smooth and intuitive that you hardly notice that you are doing it and you certainly never think of writing anything down.

If developing this designer's 'third-eye' seems too daunting, or you just find that you plain forget about it all the time, then a nice way to begin is actually to set out on observation oriented trips. Try going to a station or a major airport with the goal of finding interesting examples of how information is presented, how it is updated, how people know where things are, what colors are used for what purposes, how queuing is organized etc. As well as writing things down it is also a good idea to take pictures; the interactive world is predominantly visual, since the visual channel is a very effective route for communicating information.

Interaction takes place in many contexts. Some interactions are private and take place within small groups or just between two people. In contrast to this there are a great many interactions that take place in the public arena, and not just people interacting with other people but people interacting with things and technology in public areas. These interactions are prime material for observation, especially systems that exhibit the following:

SYSTEMS THAT ARE IN CONSTANT USE: Standing next to a ticket machine in a busy station you will see many users interact with it. Stand next to a fire alarm button and it may be many months before you see it being used.

SYSTEMS THAT HAVE A SHORT INTERACTION TIME: The more people use a system the more times you will see a complete interaction cycle. Some systems have a very short interaction cycle such as automatic doors that open and close in response to users in a matter of seconds. Other systems will have much longer cycles, consider a photo booth, the whole cycle takes several minutes. Also the more people use a system the more you will see interactions that are different or users that make errors etc.

SYSTEMS THAT HAVE A HIGH VISUAL ELEMENT: If you are observing interaction then you want something to observe. Some systems have a high degree of visually apparent material at the interface, these enable the interaction to be observed. Others have little visual material, or the visual material is hidden making it difficult to observe the interaction.

SYSTEMS THAT ARE NOT SENSITIVE IN NATURE: Even in the public arena there are some interactions that are more private than others. Always be mindful of how the user will feel having their interaction observed. This in itself is an interesting subject in interaction design, but when you are just observing interaction don't stand there looking over a persons shoulder if they are withdrawing cash from an ATM or trying to flush a urinal!

Knowledge from interacting

As well as watching others interacting you can also interact yourself. Basically, get out there and use as many things as possible. Read about new technologies, try and get hold of them or borrow them from friends. If there are two things that work in conjunction with each other try and get hold of the two of them and try them working together. Just as above there are a few points to bear in mind:

SYSTEMS THAT ARE NOT IN CONSTANT USE: In order to avoid frustrating members of the public make sure you choose a good opportunity to experiment with all the different interaction styles of a system. Choose quiet times in the system's daily cycle and be careful if long queues build up behind you.

SYSTEMS THAT DO NOT HAVE IRREVOCABLE SIDE EFFECTS: Make sure that the system is not doing something that you don't want it to. For example if you are testing a mail merging application on a computer make sure it is not sending huge mail-shots to the printer in the next room. If you are testing out the self timer on a camera make sure you are not wasting loads of film. Again, this topic of being able to experiment without irrevocable side effects is an interesting general topic in its own right.

SYSTEMS THAT ARE DIFFICULT TO WATCH: In the previous section I listed systems that would be difficult or inappropriate to observe in use by others. These are the sorts of systems that you should interact with yourself in order to get more of an idea of how they behave.

Knowledge from talking

Once you start getting more involved in technology watching you will find occasions to talk about it. There are two sorts of talking about interactive systems. Firstly there is the formal interviewing of users of systems, either in prearranged user trials or in surveys of the general public using public systems. The other sort of talking about interaction design is the chit-chat between interaction designers reminiscing about systems they have interacted with, or more interesting than that, interaction designers 'out and about' in the real world watching and analyzing things together. This sort of field trip is a great idea for anyone giving a course in interaction design. It doesn't have to be a trip to somewhere like an international airport, even a trip to the local supermarket or TV showroom is a good source of information.

Knowledge from designing

By far the best way to learn about any design discipline is to practice design. Only by actually doing design do you get to grips with the fundamentals of the area you are designing for. In the world of interaction design this entails becoming involved in commercial interactive projects, or developing your own interactive projects. Although the step of implementation of designs is an important one you can learn a lot just by designing things on paper as a hobby without ever realizing the designs. That parking ticket machine you were cursing at last week, the funny icons on the TV remote control, the web site where you think you've bought a book but haven't; sit down with a pad of paper and a coffee somewhere quiet and come up with a better design for it. You won't be able to test your designs and evaluate them, but you will have struggled however briefly with the concepts and issues that play a part in the design.

USING YOUR OBSERVATIONS

Migrating ideas to the designed world is something that I mention frequently, but how does this migration take place? To some extent a lot of it happens by 'osmosis,' simply by being aware and observing things, by building up a good appreciation of good and bad design. This will gradually have an effect on the designer's ability to design for the user.

Below are list of more concrete things to do with your observations. We can sum it up basically by saying that you ought to use those things that are good and avoid those things that are bad:

Be inspired and try to inspire

This is a less-well-definable property than some of the others we shall deal with. What I am saying is that this is something to be aware of in the real world and to try and bear in mind when designing interaction. There are examples of interaction design that can be inspirational, that make you stop and go, 'Wow – that's perfect, who thought of that?' These examples should be noted, analyzed to give general rules and those rules considered when doing your own design. But it doesn't just stop with copying the ideas, you should strive to instill that same inspiration with your own designs. Every facet should prompt the thought, 'Am I missing an opportunity to do something really good here?' You shouldn't respond by feeling that you have to do something really neat at every juncture in the design, just be aware that there are neat ways of doing things.

As an example I will cite the Quorum meeting room booking system. One feature of the design was that each meeting room had a panel outside that provided information about the reservations for that particular room. The panel had a conventional color LCD screen on it but what was interesting were the lights mounted on either end of the case holding the screen, they shone red when the room was booked and green when it was not booked. As they were on the edges of the panel they could be seen from down the corridor so you could give one quick glance up the corridor and instantly see which rooms were free.

Strive to create something with a 'feel-good' factor
There is a definite feel-good factor from good interactions. Not everybody picks up on this – for many users good interactions pass unobserved, the very fact that they are good, intuitive and natural makes them almost invisible. For the majority of users it is only the bad interactions that get noticed. Bad interactions deprive us of our ability to interact and communicate and we get a distinct 'feel-bad' factor. Lack of communication has been cited as one reason why other people's mobile phone conversations are far more annoying than other peoples face-to-face conversations; you are not fully aware of what the conversation is about and there is no possibility to interject. On a more extreme example solitary confinement as a punishment is classic removal of interaction in terms of spoken communication.

Be aware of your design responsibilities
All too often in an interaction design project it is easy to cut corners. To some extent it is easier to cut corners here than in other areas. If you cut corners in the implementation of the technology then parts of the project won't work. If you cut corners on the graphic design then things will look clumsy or unfinished. If you cut corners on the interaction design then very few people will notice since evaluation interaction design is difficult and there are very rarely concrete specifications. As a result it is easier to say, 'Oh we'll sort that out later' and not to get around to it. Functionally it works, the client is happy but the person that suffers is the end user, they are the one that must cope with inefficiencies in the interaction design, and your skipped hour or two of concentrated design problem solving can lead to thousands of lost hours as large numbers of end users struggle with the interface or have to repeat the interaction due to some problem in the design.

The best way to get an appreciation of this leverage and effect of bad design is to be on the receiving end yourself; to get frustrated at bad interfaces, to interact with things in the real world.

Avoid pitfalls
Occasionally, you come across some interaction in the real world that gets it totally and utterly wrong. Sometimes the interaction is clearly going to be bad from the onset, one look is enough for the designer to know that things are going to go wrong. At other times however you can

come across an interface that looks well thought out and complete, and only while interacting with it do you discover that there is a great failing with it.

Such bad examples need to be analyzed and the essence of what went wrong with the interaction needs to be generalized and noted down as something to be avoided at all costs.

Discover what causes annoyances

To a lesser degree there are also interactions that users find annoying. Nothing actually goes wrong, but the interaction lacks some vital feature or piece of information that assists the user in its use. Once again the lack needs to be generalized so that it can be avoided in future systems.

Consider things that are perceived negatively

As well as concrete pitfalls and less concrete annoyances, there is also the issue of things that work and seem OK but that give rise to negative perceptions for the users. The user isn't annoyed or frustrated they just feel, 'Yuk – this is not a good thing.' The best example of this are IRS (tax) forms. No one enjoys them. Even if you know that you are going to get a big refund the interaction required to assure this is still not an enjoyable one: personal questions, silly choices (single or married), tons of unnecessary questions, confusing terms, strange navigation.

I should add that my own strong feelings on this could just be a personal opinion and that in itself is an important point to remember in your observations and use of real world examples; there may be a difference in how you perceive something and how others perceive it.

Use the real world as a metaphor

There are many systems that people are used to in the real world and the act of making a designed system imitate one of these is called metaphor. A classic example is the desktop metaphor for computer systems. However, there are many metaphors that are used and many more waiting to be used, ranging from the very strong metaphors, where the system tries as much as possible to behave like the real world system it is imitating, through to the more subtle metaphors where only some fraction of the real world system is copied, for example the terminology.

This can be best illustrated by a missed opportunity in the design of PDAs. I showed the last one I was using to my kids. Morgan began scribbling and drawing with it and then, when she made a mistake, she turned the stylus around to rub her mistake out. Needless to say it didn't work, it just compounded the error, but I have seen systems since then, where reversing the stylus does turn the stylus into an erase tool.

Adopt protocols that people are used to in the real world

One example of this metaphorical copying of the real world is the protocols that people use in their interactions. In the real world people are already interacting with each other and already using interactive systems, these interactions embody protocols that those users are familiar with and that they come to expect. By copying these protocols in designed interaction you can ensure greater efficiency, fewer errors and give the user less of a feeling of alienation at the same time.

Generalize rules from interactions

With a sufficiently large collection of examples and a good abstract idea of what features are illustrated it becomes possible to extract rules from collections of real world examples. This is similar to the above point but you are not relying on the user's familiarity with the system in question, you are simply copying abstract features of the system because they work in practice.

Identify themes

As a designer gathers examples from the real world it is possible to analyze them and separate them out into different themes or categories. The patterns yielded by such an exercise can indicate interesting issues for interaction designers. Themes that have many examples are probably important ones typifying some general underlying problem that occurs in interactions. Sometimes new themes are perceived that cut across different existing ideas. These new themes can yield new ways of looking at problems.

Use the real world for testing and prototyping

Building fully-functional prototypes takes time, especially if the goal is to make them completely believable as interactive systems. It also takes time to carry out usability tests on them, no matter how casual these are. If you can find real world examples of similar interactions they can be a far easier arena to investigate, either through simple self evaluation or evaluation with two or three typical users.

Consider different ways of approaching problems

Collected examples from the real world can show the many different ways of approaching and solving key design problems.

For example, a common problem with digital media is assisting the user in choosing or finding an item from a large collection. Looking to the real world we can see many situations that can yield insights into designing to support this task. Consider the video rental store, the library, the juke-box, the garden center, the bookshop etc.

Imagine if the interface was a person

This is a classic way of illustrating the inadequacies of bad interaction design. Imagine a scenario where the same sort of interaction is exhibited not by a machine or computer but by a person. Usually the resulting behavior when couched in human terms is totally rude, stupid and unacceptable.

For example, occasionally if you make a mistake in a web form you can end up losing all the information you have just filled in. Imagine if this occurred with an interaction with a real person; 'Oh I'm sorry you've filled your date of birth out wrong. I'll just give you a new form and burn the old one so that you have to start again from scratch.'

This was the theme that I used for the Barker and Hemelsworth sketches in my book 'The User Interface: Concepts and Design' (ISBN 0954723902). A Lord in his mansion served by a well-meaning butler who operated in the same way as a badly designed computer system. The behavior is often familiar to computer users and totally unacceptable when exhibited by a person no matter how well-meaning they are.

Adopt real world terminology

A simple error of interface that is often made is not to consider the terminology of the interface. The visual design and the behavior have to be designed, that is obvious. However in discussions between the design team certain terms are used to discuss the design and all too often these can end up being used in the interface without being properly considered. My favorite example is the 'mark' introduced in the MiniDisc recorder on page 112.

Finally then, on to the columns, by all means use these as a source for ideas and material but they are most valuable if they are viewed as an example of how to go about collecting a store of anecdotes and ideas for yourself.

THE COLUMNS

part one

H A R D W A R E

One of the things that inhabit the world along with us humans are devices; technological things that do things for us: telephones, ATMs, video recorders, the list is endless. Just look around yourself now and you will see what I mean. It really is incredible, not only how many of these gadgets surround us, but how we have got so used to them being there that we are almost blind to them. As far as the interaction designer is concerned these devices are the front line of designing for the user. They are the day-to-day 'skin' between the world of technology and the world of the human.

The columns in this part of the book start by looking at the ingredients of the interfaces to these devices, the parts of the interface that the user interacts with: the buttons and slider controls. The focus then moves deeper into the interface to look at the underlying functions behind those controls: powerful functions, safety catches. These first two sections in effect deal with the syntax and the semantics of the interface to devices. The syntax are the buttons, sliders etc that the user encounters at the interface. The semantics are the functions; what happens when the buttons and sliders are activated. The semantics are in effect what the buttons 'mean'. The terms 'syntax' and 'semantics' are actually terms concerned with language. The syntax of a language are the words and how they fit together while the semantics are what those words mean.

The final group of columns here cover more concrete examples of devices: elevators, coffee machines and the like. This is the 'bread and butter' of interaction design; analytical consideration of the devices that we interact with on a day-to-day basis at home or at work. We will be considering things as simple as elevators and things as complex as video-conferencing. There are also a few examples that are taken from a wider scope than devices, in particular the interaction design aspects of cooking (cooking in a restaurant is a task as precise and deadline-bound as any military campaign).

Buttons

'**62**...*63...64...*' What am I doing? '*65...66...67...*' Counting sheep? No. '*68...69...70...*' No such luck. I'm at work. '*71...72...73...74...75...*' New printer. '*76...77...78...79...*' You can attach it to the Ethernet and send the print jobs to it over the net. Brilliant! '*80...81...82...* ' But first you have to program the IP number in, (the Ethernet address). '*83...84...85...*' There's a clever menu structure, you use one button to step through the menus and another to step through the choices in each menu '*86...87...88...*' and a third button to select the menu option and do it. '*89...90...91...*' But for the Ethernet address you then have to program the digits in and there isn't a numeric key pad. For each byte in the address you have to start from 1 and then press the next button to get 2 and then next again and so on. '*93...94...95...*' The address is 195.78.77.120. Thank goodness I only have to do it once!

'*15...16...17...*' Now what am I doing? No, it's not the second byte. I finished them all. It took ages! So what am I doing now? '*18...19...20...*' Well it turns out that when you've reached the number you want for the byte, you have to press the enter button before going on to the next, otherwise it ignores your changes. '*21...22...23...*' So I stepped through setting each byte and when I'd finished them I checked the address and it was 000.00.00.000! When I have finally finished (assuming nothing else goes wrong), I'll have pressed that stupid button about a thousand times, I can already feel RSI setting in...

In the hi-fi world there is a trend among manufacturers for as many knobs and buttons as possible. In other sectors of the electronics market they try and get by with as few buttons as possible. How often have you programmed the time into a digital alarm-clock where there is a fast forward button but no reverse button? The time whizzes past, and whoops! You miss the setting you wanted and have to go right round the clock again.

Less buttons is not, by definition, a bad idea. If a system is well designed and comprehensive then a small number of buttons can be wonderful. The key phrase there is well designed, but most companies are not motivated by good design. The drive towards as few buttons as possible is motivated by the cost of building buttons into equipment, the lack of extra space for buttons and labels and the increased maintenance and decreased life-time that come with extra moving parts. It certainly has little to do with a drive towards simplicity and ease of use! Indeed the lack of good design means that button deficiency leads to several detrimental effects:

1) Some actions take a long time due to using a button with a general simple functionality to achieve a complex goal.
2) Some actions do not have a suitable undo like the alarm-clock mentioned above.
3) Some buttons have different actions in different contexts (modes) leading to errors. Don Norman cites a digital watch where the button to illuminate the display also functions (in stopwatch mode) as a reset button. A boggling nightmare for night-time joggers!

An interesting side-effect of button limitations is that as well as cramming all the user functions into a small set of buttons the designers also have to cram the maintenance, testing and demo functions into the same set, hiding them in strange multi-key combinations. Try this at home, or at work; select a convenient (and non mission-critical) piece of technology with buttons and try pressing different combinations of buttons down at the same time. When I tried this with my digital watch I found a frightening demo mode with flashing patterns and a mindlessly silly tune, on a fax machine I discovered hidden maintenance and configuration modes. I try it regularly with the chocolate machine while I'm waiting for the train, but have yet to find the 'free chocolate' mode.

In the world of computers button limitations govern keyboard commands and mouse-button functions. But there are also soft, on-screen buttons. Give a programmer with a graphic toolkit two minutes and they can knock together a screen with hundreds of 3D-looking buttons. Add cascading menus and button panels and you have a couple of hundred more.

But this incredible freedom is not necessarily good. In this new, soft world there should be limits to the number of buttons, but now instead of being constrained by cost and size there are other limits that should play a role. These limits are bound up less with the physical world, and more with the mental world of the user. The number of buttons should be limited by the functionality of the system, by a concise yet comprehensive user model for the systems and by opting for, and clearly supporting, good modality choices. Physical constraints are easy to deal with; three buttons take up more space and cost more than two buttons, but these new limitations, coming as they do from the human side of the interface, are more amorphous, more complex and infinitely more interesting.

Right, now that I've finished I'll try and print this out on our new network printer. Just a click on the button and...

Slider Controls

A friend of mine almost succeeded in blowing me up last time she visited us. It was accidental of course, and it is directly connected to user interface design. It all has to do with the knobs for controlling the gas burners on our cooker. In Britain turning the knob anti-clockwise takes the burner from the 'off' state to 'on-low' (simmering) and turning it further increases the gas making the flames hotter until at the end the burner is on full. This is a similar situation to combined on/off volume controls on radios; you turn the dial taking it from 'off' to 'on-low' and through to 'on-high'.

Contrast this to gas knobs in some other European countries where the sequence is, 'off' to 'on-full', and then down to 'on-low'; the simmering state is right at the end of the range instead of at the beginning.

This has the advantage that you can quickly turn the knob into the simmer state while cooking, without the danger of it slipping into the off state if you go to far. The disadvantage is that due to the inconsistencies in the two systems, visitors from Britain trying to turn the gas off will often turn the knob from 'on-full' to 'simmer' and because the knob won't turn any further they then assume it is off. If it is lit it remains burning on simmer, but if for some reason it wasn't lit then it remains pumping a small stream of gas into the kitchen…

In the column on page 88 I discuss snooze functions; states that sit in between two distinct states. Another important concept are states that lie in between two other states but they are all part of a range. The simplest example is the volume knob mentioned above.

Often with such systems there are irregularities introduced in the interaction to isolate key states in the range. The most obvious states are the beginning state and the end state. In some interactions these states are desirable for some reason and so are made difficult to leave. Consider again the gas knob; you usually have to push it in before you turn it on, the state at the beginning of the range (off) is isolated and given a threshold from the rest of the range.

Other examples are car doors; when a door is fully closed it is useful to have it remain in this state while traveling, thus doors have catches to make getting out of this state a threshold action (pretty common sense actually).

Key states in such ranges are not just given a threshold in this way, sometimes the system has a built in tendency to return to the state; doors that tend to close themselves, joysticks that auto-matically re-center themselves, etc.

The interesting group is ranges where the key state is not at the end or beginning but is an interim state; somewhere in the middle. Consider car doors again; you are very often opening them in cramped situations, parked next to other cars, parked alongside a busy road. In those situations you may want the door open but not open all the way. Many car doors support this by having an interim state built in, a point at which the door can be left half open and it stays half open.

Similar interim states can be found on some hi-fi controls. The balance knob usually has a little tactile 'bump' when the balance is equal for the two channels. Some British cookers pick out the simmer state in the range in a similar way.

These interim states are built in, based upon objective ideas as to which states are important to the users. But very often with such ranges the important state is purely subjective and as such cannot be built in. With our electric toaster, the cooking time control has one state which is vital to me; the state for nicely done toast. However this is different to the equivalent state for my girlfriend, sometimes leading to explosions of a different kind in the kitchen!

The only thing that systems can offer to help here are presets; several independent controls for the range and a way of switching between them. Televisions supply you with up to a hundred presets, imagine zapping if you just had one dial like a radio. So, toaster manufacturers are you listening; we'd have a range control for me (that I could set to nicely done toast), and another one for Wendelynne (that she could set to warm, floppy, slightly-singed bread), oh, and a third one for whoever is visiting from Britain. That is if they hadn't already blown the kitchen up.

Volume Control

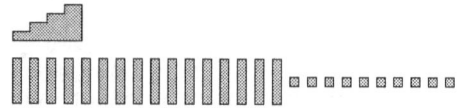

Sometimes, before I go to bed, I have a quick look at the news headlines on the TV. I switch it on and, 'Boom!' It comes on with full volume, waking up the kids, making me jump, even making the cat lift a sleepy head. I have got into the habit of switching it on and straight away hammering on the button to lower the volume so that when it comes on the volume is at least on the way down. This is yet another example of those great technology 'fear and panic' moments; will the heating really come on at six? Has the back-up really worked? Why are the numbers flashing? What's it beeping for? Recently though, I discovered a different way round the problem. I turn the TV on, immediately hit the mute button so that it comes on with no sound. Then I re-tune it to the video channel which has no signal. Here I can adjust the volume (when you adjust the volume it takes the mute off so you can hear what you are doing) in comfort. Then I switch back to the normal TV channels and the sound is at an acceptable level. Only an interaction designer can have more fun switching the TV on than actually watching it!

In another column here I discuss the use of controls to alter a value within a range, like the volume level, gas cooker controls, light dimmers etc. Here I am concentrating purely on controlling volume. Those TV controls sum it up, there is a control to alter the level and a control to switch the mute on and off. These two controls (or three if you have separate controls for volume up and volume down) seem clear-cut, but there are all sorts of hidden complexities that need to be designed and not just left to the whim of the technicians putting the system together.

For starters, there is the issue alluded to above concerning how the controls interact. When you mute the sound and then alter the volume should it come back on or stay muted until you un-mute it? Coming on automatically means that you can hear what you are doing, it also catches the situation where a user gets confused because there is no sound and tries to remedy the situation by turning the volume up. If the volume control does not automatically turn the volume up then the user could end up putting the volume to maximum before trying the mute button resulting in them getting their head blown off when the sound comes on again. Another important issue is what happens to the values when the system is switched off and then switched back on again. Does mute stay on? Does the volume stay at the level it was at when the set was switched off? In the physical world the volume dial usually just stays set on the same volume. Only if the volume dial includes the on/off switch will the act of switching if off influence the volume. The advantage to a combined on/off volume dial is that whenever you switch it on it comes on quietly, the disadvantage is that if you are fussy about the exact volume you are listening at then you have to reset it each time.

Presenting the controls like this is not much of a problem when you are dealing with physical controls on a product, their presentation is governed by their physical form. Presentation can become a problem when you are presenting things as on-screen controls in a digital context. On-screen controls don't cost anything to add, so if you have a computer with a complex set of sound-card utilities the chances are that there are a good number of volume sliders dotted around in different control panels, each of which performs exactly the same function. I can remember spending ages adjusting the sound-card volume slider and the PC volume control before realizing that they were just two views of the same thing. Can you imagine a physical system that somehow has several different controls that all do the same thing?

As well as being able to present it in different places on screen, there are also different ways of doing the actual presentation. The most common form of the sound controls is a slider for volume and a little checkbox (akin to an on/off switch) for the mute. The user can either hear what is going on or, in the event that they can't hear anything, they can see what is going on. If space is scarce these controls can be hidden in a pop-up control panel but when that is not visible it is still good practice to give some visual indication of the state of the mute control. Some digital systems save space by just having three little controls: a mute and an up/down for the volume. But with such a set up you can see if the mute is on or off but you can't see the level of the volume. Also not showing the level as a drag-able slider leads to confusion about the cooperation of the controls; if I mute the sound and then do 'volume down' am I having no effect because the sound is on mute or am I changing the volume level so that when it comes back on again it will be at the new, lower volume?

On my laptop there is very strange visual feedback. Adjusting the volume with the volume up and down buttons calls up a row of green marks on the bottom of the screen showing the level. When I hit the mute button however, the indicator shows these marks fading away, but it seems to do this just as a consequence of hitting the mute button, even when it is already muted every time you hit the mute button it shows these green marks and then fades them away, initially I used to think 'Gosh, good job I muted it because the volume was on quite high'; it was only after several months that I realized that it was usually already on mute anyway. OK it is a small detail but it does bring that twinge of doubt into your mind about what you think is going on and what the computer thinks is going on.

Trust is a huge issue but it is built up from very small building blocks. Volume control is one of those basic functions that appear in all sorts of designed systems. Most designed solutions to the problem do work even if they do have their idiosyncrasies, but like the laptop example if you can't trust your computer one hundred percent on a simple job like adjusting the volume then how can you trust it with a job like performing an integrated email marketing campaign?

Off and On

I have an electric water-kettle with a little red light on it. The interaction design is enough to make your hair curl; the light goes on when the kettle is plugged in and it goes off when you press the switch to boil the water. Straight away there are problems because the kettle is using one light for three states: 'off', 'on' and 'not-plugged-in-you-git'. Furthermore, the two states that are furthest apart in terms of what the user wants are the two states that are represented with the light off. Namely the state of: 'kettle is on, heating up and everything is just fine, you will be drinking hot coffee in just a few minutes', and the state of: 'kettle doesn't even have any power you idiot, it will be ages before it starts boiling because that will never happen. In about ten minutes you will wonder why you are not drinking a cup of hot coffee and then you'll figure it out.'

I also used to have a toaster that solved the 'not-plugged-in-you-git' problem by not allowing the user to keep the toast handle pressed down when there was no power. You put a slice of bread in, pressed the handle down, the toast holder slides down into the machine, you let go and instead of staying in there, the bread just popped up straight away. There can be problems with misinterpretation of this, I have watched visitors to the house repeatedly banging away at the handle to press the bread down thinking that the toaster was a bit old and didn't stay down properly. Understandably they didn't immediately think to check the power supply.

Computers too have on/off indicators. Sometimes they are a simple red LED, but sometimes they are more involved. There are good designs and bad designs. Early Sun computers, (and I mean 'early' here; pre-IBM PC, when PERQ was just a gleam in Sun's eye) had an awful design. A Sun workstation in that time was an industrial-strength collection of gray units connected by thick, coiled cables. The on/off indicator was a row of red LEDs next to the power socket on the back of the system box. To show that the computer had power these LEDs would be flashed in sequence from left to right and then again from right to left, the effect was of a little bright red dot zipping from side to side continually. Some technician somewhere had obviously said, 'Hey, if we put a line of LEDs in we could do this it would look really cool.' Fortunately it was tucked away on the back of the unit. Unfortunately these units were often placed back-to-back in shared office spaces so that the user could see the LEDs on the back of the unit opposite, with the result that their gaze was being continually distracted by whizzing bright, red lights in their peripheral vision. Much later, a Silicon Graphics machine (was it the Indigo?) had a different approach; one LED that indicated the machine's status by its color: red = power on, yellow = booting up, green = ready, black = 'not-plugged-in-you-git'. In my next book ('Design for New Media' Addison-Wesley 2004, ISBN 0201-596091) I mention the standby-mode indication in the Apple G4s when they were first introduced. It was on/off with an edge; the light 'pulsed' on and off like something from a 50s sci-fi movie. The communication was both functional and emotional. Not just 'I am on and in standby mode' but 'I am super powerful and am waiting like a crouching tiger for your next command.'

By far the most annoying 'is it on?' syndrome occurs when setting things up to switch on in the future, things like video recorders and alarm-clocks. Especially unfamiliar and cheap digital alarm-clocks in hotel rooms. Is it going to wake me up at 7:30 so I can finish my presentation slides or not? As well as problems with things being on or off in the future there are also usability problems with things having being on or off in the past. Heat falls into this category; the clothes-iron switched off and put in the cupboard, the electric ring on the cooker that was on a minute ago. The indicator light is off, but that is not the important fact as far as the user is concerned, the user should be told if it is hot or not and there is not a one-to-one relationship between being on and being hot. This problem is elegantly solved by the Braun hair-curler which has a nice heat-sensitive blob on the end that changes color according to whether it is hot or cold, irrespective of power being on or off. Now that really is a bit of interaction design guaranteed to make your hair curl!

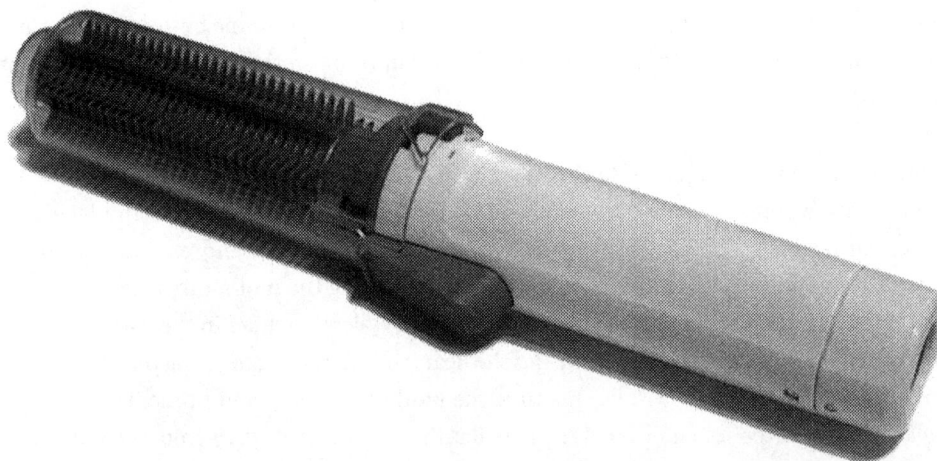

Powerful Functions

Imagine that you have a thousand monkeys sitting behind a thousand computer keyboards and that you set them tapping at random at the letters. It is said that if they carry on doing this day in day out eventually, after many years, one of them will just happen to type out the entire works of Shakespeare. It is also said that if you leave them tapping away for just six months then more than one of them will write a buggy, but marketable computer operating system.

Imagine how much faster it would be if the keyboards they were working on had one key mapped to a macro that produced all the ASCII for the complete works of Shakespeare; a real powerful function. After lining all the monkeys up you would only have to set them going for about five minutes before one of them hit the button and the task was complete.

Nowadays more and more technical systems are being equipped with these sort of powerful functions; one button that carries out a complete task. While this may be useful for meeting literary deadlines with a thousand monkeys, it does have disadvantages when it comes to error robustness. On any technological system a button can be pressed by accident by the user, by another party or just by being poked by something else in your rucksack. The effect of such an accidental poke is governed by the power of the function. The more powerful the function, the more serious the result.

At home I often let my two-year-old play with our, rather outdated, dial telephone, safe in the knowledge that no matter how much she pokes and prods the chances that she will dial the Chinese speaking clock and then leave the telephone off the hook are astronomically small. However, when I am on the phone I keep well away from her since the button for hanging up then becomes a powerful function, a simple poke can result in the interruption of the entire conversation.

While we are on the subject of telephones there is the fascinating 'dial M for Murder' story reported in the press a while back. A woman answers a phone call at home and hears, not a voice, but obvious sounds of a struggle. Then she hears a scream and realizes that it is her daughter screaming. Suddenly the phone goes dead leaving the mother in a state of shock. Is her daughter being kidnapped, or has she disturbed a violent intruder in the flat? Her fears are confirmed when the phone goes again and she hears the same struggle going on and her daughter's voice shouting 'oh my God'. This time the mother hangs up and immediately alerts the police. When the police enter the daughters flat they are met by a very embarrassed daughter and her boy-friend and it emerges that in the frantic throes of their lovemaking someone's foot knocked the receiver off the phone next to the bed and poked the 'last number redial' button. A simple accident, but serious results due to the power of the function under the button.

So, power functions incorporate efficiency, but the downside is that they cause problems if activated in error.

The negative aspects of powerful functions do not just involve accidental usage. Some powerful functions have bad side-effects from everyday use. Consider the 'copy file' function. In days gone by when everything was written by pen and typewriter the concept of a back-up copy didn't really exist. I remember the advert in a student magazine in Manchester:

Stolen from the physics department car park; one light blue Ford Fiesta. You can keep the car but please return the doctoral thesis that was on the back seat, it's my only copy and I have to submit next month!

Nowadays, with the powerful copy/duplicate commands, creating a copy is simply a menu option away. Many editors make back-up copies automatically. The result is that if you are paranoid about computer crashes (as every computer user should be!) then your hard disk, the hard disk at home, the local hard disk and the email box of the place where you used to work and still have an account are full of different versions of the same document. You make copies as easily as breathing. You want someone to proofread it, you make a copy for them, you want to change the layout, you make a copy, convert it to ASCII, you make a copy. The result is a stagnant file system full of redundant copies where finding the last one that you worked on is as difficult as trying to trace a stolen Ford Fiesta.

Power functions also play a role in creative applications. Consider cameras with built in exposure programs and sets of filters. The glitzy results are easy to achieve, but the problem is that the results look as if they have been achieved simply by using the right combination of filters. This observation is also true but to a lesser extent for the Photoshop artwork software. In the creative world power functions once again make the difficult easier, but here the downside is that they can limit the creativity.

Computers have many more keys to be poked than a telephone and software is far more complex than a phone system thus amplifying the above problems. Among the many software benchmarks is a little-known test for software stability called 'the open book test'. You install the software, set it running and then open a large, heavy book and rest it on the keyboard as though to read it...

Safety Catches

It's the climax of a science fiction movie. The all-powerful-beings-with-the-funny-heads have got the spaceship in their clutches and have ordered the captain to send his crew down to the planet's surface. The captain tells them that he will make the preparations. He breaks contact and then turns to the second in command. 'Anything is better than that, prepare to self-destruct the spaceship.' The two of them flip the protective flaps on their control panels and both press the finger-print activated buttons together. The spaceship explodes in a huge fireball leaving the all-powerful-beings-with-the-funny-heads confused and angry.

Sounds familiar? Of course it doesn't sound familiar! Self-destructing a spaceship is never that easy. What really happens is that the captain has to ask someone in engineering to initiate the self-destruct sequence and then it takes several long minutes with spoken countdowns, dimming lights and in the final stages plenty of sparks and shuddering. A delay that is not very efficient if you want to blow the spaceship up quickly, but that is vital in that it gives the all-powerful-beings-with-the-funny-heads time to say 'We don't understand, you humans would rather destroy yourselves than submit to the wishes of us all-powerful-beings-with-the-funny-heads. We need time to think about this, you are free to go.'

In the last column I wrote about powerful functions and what happened when they were used in error. Well, many powerful functions are so powerful that their use needs to be made artificially complex so that accidental use is minimized. Think of the safety catch on a fire extinguisher. First the user has to remove the safety catch and then they can use the powerful function. Usually the more powerful the function, the more safety catches it has on it, or the greater the number of discrete actions that need to be taken to activate the function.

The system has a state in the interaction that is powerful and therefore dangerous in certain circumstances. Access to this state in the

only press this button if you're really sure... **self destruct**

interaction is blocked by one or more buffer states, states that have no other function than to make the route to the powerful state more tricky and thus less likely to be reached by accident. Getting back to science fiction movies again, buffer states always play an important role in self-destruction sequences: blowing up the mother ship in 'Alien'; 'Star Trek' (where it almost seems to be standard procedure on first contact) and '2001' where they behave slightly more seriously, no big self-destruct scenes but well designed buffer zones for other systems. Think of Dave preparing to blow the explosive bolts on the pod door (five or six distinct actions accompanied by five or six distinct alarm sounds) or the final shutdown of the HAL computer, a long drawn out affair with dimming lights and slurred speech. Apart from the shutting down of HAL such buffer states also exist in more mundane computer operations such as deleting documents on your PC: The eternal 'Are you sure you want to do this?' dialogue box. Actually I think that during that scene in 2001 HAL does say 'Are you sure you want to do this Dave?'

With some of the early word processors even the act of saving one file over another was fraught with 'Are you sure? [y or n]' questions leading to the 'Y file syndrome'. Users would sit there hammering on the 'y' key in response to all the silly confirmation questions only to discover that they were through to the other side of the dialogue and somewhere along the line they had responded to the question of 'What is the file to be called' also with a 'y'. (And now that I think of it 'the Y file syndrome' would make a good title for another science fiction movie.)

With real world systems there is a balance to be struck between two things; the state that is being protected has dangerous consequences but it also plays a vital safety function. These states need to be protected with buffer states, but they also need to be quickly and clearly accessible in an emergency. Think of activating fire extinguishers, the emergency brakes on a train or opening the emergency door on an airplane.

The key goal with safety catches is that only intentional use is possible and when the user intends to reach the state it can be reached simply and quickly.

Finally, talking about all-powerful-beings-with-funny-heads readers may be interested to know that at the time of writing the British elections were in progress. After the major parties had released their manifestos, Labor's manifesto was so simple and clear that the Labor leadership publicly described it as a 'What You See Is What You Get Manifesto'. User interface design comes of age!

Coffee

Programs run on computers. Programmers run on coffee. Indeed I've heard it said that CASE should really be an acronym for Coffee Aided Software Engineering, and if you are an international researcher you come across many different sorts of coffee, ranging from the thimble-full of super concentrate that you have to gulp down quickly before it eats its way through the cup to strange concoctions where you would rather eat your way through the cup than have to drink what's in it.

Anyway, an addition to the opening phrase about coffee could be that user interface designers run on coffee machines. At any gathering revolving around user interface design huge discussions arise while interacting with everything from the light switches to the vending machines. A few months ago I was giving a series of lectures on simple user interface design. During the day drinks breaks were scheduled in and we would gather round the incredibly complex vending machine, where the topic that I had just been dealing with would be brilliantly illustrated, usually by the bad design of the machine. It incorporated hot drinks, chocolate bars and LCD screens providing huge amounts of information to the user and it all worked with cashcards. There was a built in machine where you could buy a cashcard and feed coins in to have your cashcard charged with money, or you could return your cashcard and get the leftover cash back.

One of the subjects I covered in the talk was feedback. During the coffee break we discussed how the machine was a corporate model, meaning that it looked like a corporate headquarters; smooth, smoked glass unbroken by buttons or text. The controls were touch sensitive areas hardly distinguishable from the rest of the machine. In the afternoon session I explained that too much unnecessary feedback was a bad thing. In the break we saw that the LCD screen of the vending machine supplied screen-full after screen-full for every key press. No wonder the queues moved so slowly.

I talked about user configuration of systems, how it should be simple. During the coffee break one of the researchers who worked there told me that the cards could be programmed with your choice so that all you had to do was stick the card in and it would give you the drink and deduct the cash from your card. He also said that when the machines were introduced the staff had to have a three week course in using them. The machines were so complex that I didn't realize he was joking.

Funnily enough this mention of programming the cards explained why the cashcard I had bought from the machine always insisted on giving me a cappuccino. The previous user had programmed it and somehow the programming had remained on it for the next confused recipient; me! It was lucky that cappuccino was also my choice since I had no idea how to reprogram it. This observation tied nicely in with the section of my talk about the problems of building defaults into interactive systems.

Also during the talk I dealt with the disadvantages of designing an interactive system so that the user has to perform many actions in order to reach a particular goal. Later, in the coffee break, I tried to buy a bar of chocolate from the vending machine. The chocolate was all on show, arranged in a grid behind glass, and each choice was numbered according to the row and column it was in, thus a Mars bar was 25 (row 2 column 5), a Bounty was 26 and so on. The interaction actually involved to get a Mars bar was more complex. I first had to type a 2 in to select the row and then the LCD screen told me what was in that row (even though it was plainly visible behind the glass). I could then type the second number in to select from that row. Fine, selection made, I thought. But no! Once I had actually chosen the chocolate bar, the LCD screen informed me that I then had to press the START button to actually get the machine to give the chocolate to me; what else was it expecting me to do with the choice I had made, delete it? Move it? Three interactions to make one choice – not good.

Clever user interface designers will now be pointing out that I didn't need to deal with the actual choice as two digits, if I had just keyed 25 in as a two digit group I would have chosen a Mars bar in only one step and been unaware of the fact that the choice was made up of two actions. However, I thought of this at the time and tried it, only to discover that after typing in the first digit there was a short 'dead-time' while the machine processed this and altered the LCD screen. During this 'dead-time' it didn't register the second digit. Not good at all.

The presence of that particular vending machine next to the room where I was giving the course was quite a coincidence, but even without coincidences our daily lives are strewn with interactive products, everything from cat-flaps to CADCAM systems. A good user interface designer is always a good user interface designer, even when they are making a phone call, interacting with a ticket machine or buying a cup of coffee from a hot drinks vending machine. Even more so if they teach user interface design, they should always be filing away examples to be used to illustrate good or bad user interface design in a way that everyone can appreciate and empathize with. So the moral of the story is: keep your eyes and ears open in the real world, not just while you are at work designing or researching user interfaces, there's a lot to learn out there. And funnily enough, now that I think about it, that machine may have had a complex and confusing user interface, but it did actually make a good cup of coffee!

Telephones

As the telecom world spirals upwards, out of control, the humble consumer is being bombarded with no end of new and wonderful telecommunication services. What I am talking about here are not the sub-micro, GPS-enabled, satellite, internet, mobile telephones that people use to shout 'Hi, I'm on the train, I'm going to be about ten minutes late.' I'm talking about the new interactive services that are being produced by clever (and not so clever) combinations of technological possibilities. Digital sound storage, knowing the number of incoming calls, touch-tone feedback.

By ringing a certain service number you can have an automated voice tell you the telephone number of the last person that called you. Great! We would come home after a day in the park and ring the service number only to spend the next ten minutes puzzling over the number that had last rang us. '9733445 do you know who that is?'. 'Well it can't be John.' Several times we ended up ringing the offending number just to satisfy our curiosity. 'Hello … no, I don't know why I'm ringing, why did you ring us?'

Then there is the voice mail system. We have moved into a new house, sorted out phone and electricity but it took us two weeks to discover that we also had voice mail. I just happened to ring home from work and stay on the line long enough for it to cut in. When we sorted out access codes that evening there was a pile up of message from people built up over the last two weeks, the last of which was me saying 'Ohhh! Gosh. I didn't know we had voice mail.'

The frustrating thing with the voice mail service and with most other telephone services is the long drawn out nature of the interaction. The first few times around it is useful to be spoken too like a deaf simpleton, but after a few months of daily use it gets tedious when each interaction begins with 'welcome to the voice mail messaging service.' Being greeted each day anew is something I expect from my loved ones not from my telephone.

Worse still the voice mail service is 'intelligent'. Unlike a simple answerphone it will take a message from a caller who rings while you are engaged in another conversation. Indeed the voice mail is so intelligent that it even goes so far as to ring you up after you have completed the call to tell you that there is a message. Useful, but it was quite a shock to have the dormant, subservient voice mail lady suddenly taking things into her own hands and ringing me up for a change. 'Wait a minute, you can't call me, I'm meant to call you!'

However the strangest service was two jobs ago. Working late in the lab. No one was at home so I had set up the home telephone to pass all calls through to my work number (I used to do this a lot until I found out how much it cost)! It was late and dark, there was no one there to talk to so during a break in the writing I idly wondered what would happen if I rang my home number from work. What sort of infinite loop would the phone system attempt to set up as I rang home and my home telephone tried to route it back to me? Would I be able to hear myself? In the deserted and eerie lab I rang my deserted home. There was a few seconds beeping... and then a deep, male voice said 'yes, hello?!' Who was it? A burglar at home? The secret service tapping the line? Some strange, lost spirit that had got stuck in the telephone exchange and only got a chance to talk to humans late at night when they tried strange things with the telephone net?

No. It turned out to be the night porter. After-hours he functions as the telephone switchboard as well. When I got redirected back to my own number it was engaged (by me!) and so the call was passed on to him. However, despite this simple logic we did have a very strange start to our conversation as he first had to convince me that he wasn't some strange lost spirit that had got stuck in the telephone exchange.

Elevators

Have you ever noticed that when you are watching a thriller and it reaches its nail-biting peak, the final scene is always played out in a dark, deserted and spooky theater, a dark, deserted and spooky underground tunnel or in an ordinary elevator.

What is it about elevators? Why is an ordinary elevator so scary? Why aren't other interactive systems like coffee machines or video recorders frightening in the same way? It could be that they incorporate the idea of uncontrollable interactive technology ('Open the pod-bay doors HAL!') or that the interaction actually involves the user getting transported somewhere.

Anyway, scary or not, elevators are complex, highly-interactive, multi-user systems and are full of good illustrations of user interface design. One nice example that I often quote is the combination of audio and visual feedback used when you are waiting at a group of three or four elevators. Audio feedback is intrusive and powerful, it is good for getting attention and alerting users to changes in the system. Visual feedback is more passive than audio, but it is far more context oriented, it has a spatial element and it can be used to label things and positions. When a elevator arrives the 'ding-dong' tone alerts you to the fact that something has happened and the lit arrows above the elevator doors shows you which elevator has arrived. This makes good use of the strengths of each medium. Such a feedback combination is also used on a desk full of phones where the phones have flashing lights on them so you can see which one is ringing. (A question of timing; in what order should you use the two feedback channels)?

An elevator is not just a source of interesting examples of interaction, there are also examples of information presentation. Consider the elevator I saw where the buttons for the floors were ordered in two columns to save space on the console, the floor indicator digits were arranged in a horizontal row left-to-right above the door and the list of companies on each floor was done in the conventional order with floor one at the top and floor twelve at the bottom. Three different presentations of the same information, none of which had an accurate relationship to the actual arrangement of the floors!

Then there is the eternal numbering problem. The Europeans number their floors starting from the bottom with the 'ground floor' while the Americans start from the bottom with the 'first floor'. Further complications arise in European lands where the word for ground floor begins with the letter B and the buttons read '3, 2, 1, B'. The letter 'B' can be read as 'basement' thus supporting the (in this case incorrect) user model that the American numbering system applies. An interesting solution to this dilemma in floor labeling was used in a tower block for

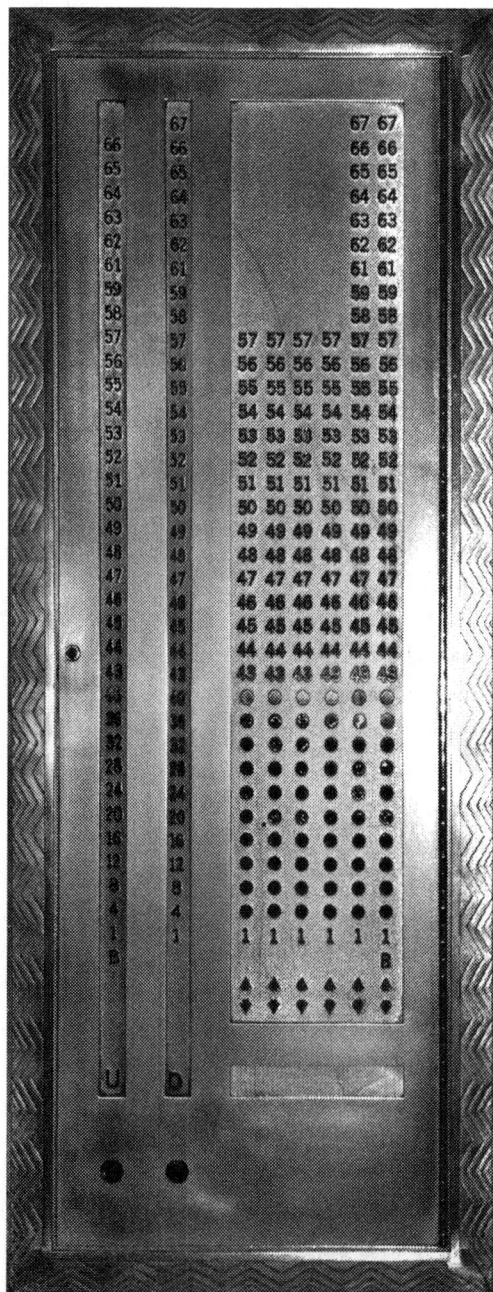

a particular university math department. It was built in the 60s when there were lots of new and strange ideas about architecture and design. The designers had chosen to use the letters of the alphabet to label the floors. This did have the advantage that the math department could be on floor M and computing on floor C and so on, but the weird thing was that they had decided not to include the vowels. Was this to avoid the confusion between the letters I and O and the digits one and zero? I wish I knew. If the architect is reading this will he or she please drop me a line.

However, the most disconcerting of all the elevator related information I have seen in an elevator was in an international airport. A clearly printed sign in the elevator listed the floors and the facilities available on them with an equally clearly printed declaration 'you are here' next to floor 2!

To close I should say that you do get the occasional, real life, elevator horror story. A colleague of mine once lived in an apartment block serviced by a secure, key-operated elevator that opened directly into the apartments. If you took the elevator down to the bottom, the journey was not always as direct as it should be. Sometimes, by some quirk of elevator logic, it would stop and open the doors in the third floor apartment, the couple were always out, but over on the other side of the huge, dark living room, their huge, dark dog would suddenly catch sight of you…

Video

In the light of the Starr Inquiry and the impeachment maybe I too should make a confession; I have been involved in user interface design for fourteen years and only in the last two weeks have I acquired my first video recorder.

It's amazing, it has changed my life completely; I don't have to wrack my brains for hours to write an article about usability, I can start conversations with other user interface specialists without having to wait until we are in an elevator and, instead of spending all my evenings at cafés and cinemas (and changing diapers), I spend them at home with my family gathered around the TV trying to fathom out the workings of the system.

Take the 'energy saving' feature; I was trying to get the hang of programming the video and set it to record in an hour's time. While waiting we settled down to watch 'The Fugitive' with Harrison Ford on TV, a real thriller; falsely arrested, imprisoned, caught up in a jail-break from the prison van, it crashes onto a railway line and then just as the train came hooting into view... the picture went blank. The video had switched itself off taking the TV picture with it. I switched it quickly back on. A few minutes later the same thing happened. It seemed to be happening at intervals of about 10 minutes. What was the machine assuming I was doing, what was its task model of me? Did the user usually program the video to record and then go to bed thus making it a good idea for the video to switch itself off automatically? Energy saving? The amount of mental energy I wasted struggling with the feature far outweighed any electrical energy that it may have saved.

Another example of advanced features occurred when we got fed up with 'The Doctor' with William Hurt. Halfway through we decided to stop the video, catch 'The X Files' on TV, and watch the rest of the video the next evening. Twenty minutes later, while I was making a night-time drink for myself, I suddenly felt as though I was starring in an episode of 'The X Files' myself. From the kitchen I heard the video machine reactivate itself and start whirring, what was it doing now? As I got to the doorway it ejected the videocassette which had been rewound right back to the beginning. An incredibly useful feature if you want to stop a movie halfway through and rewind the cassette right back to the beginning, but why would you want to do that?

What was the designers model of my task? What was the goal I was meant to be aiming towards that this machine was helping me with? How long did it give me before it started rewinding? Would I have time in future to stop the tape without being rushed back to the beginning of the movie? Basically what I needed was not a set of instructions to tell me how to make the recorder fit my model of what it should do but a set of instructions telling me how to adapt my life so that it better matched the complex task model that the video was based upon.

As an example of simple faults consider the disconcerting readout on the front panel, under normal operations this always showed the message 'E1'. Now, anybody with experience of compilers for memory scarce 70s computers will interpret this as a curt compiler message something like: 'E1: Error one, syntax error, correct the syntax problem and continue programming.' Having this huge error message gazing at me at all times was worrying, I would have been happier if it was showing something simple and reassuring like 'OK' or 'A1'.

Analyzing all these features has made me look at our trusty old TV through new eyes and I realize that it too has some awful attributes. On the control panel at the front of the TV are two vital buttons, one allows you to tune in the channel you are currently watching (so you can zap through the channels with the remote and then tune each one in) the other is an automatic 'tune all channels in' function that allocates each number to channels in their order on the waveband; it always seems to result in channel 0 to 9 being Arabic or Turkish for some reason. Anyway, these two buttons are identically colored, are about the size of split peas and are mounted very close together and about half an inch off the ground. And guess what? Yes, sometimes I go to tune in a channel and I press the wrong one thus losing all my hours work of presets in a split second.

And another thing! The power-on LED is miles away from the actual on/off button and both of them are mounted at floor level. When we turn in for the night I switch the lights out and then home in on the red LED and start kicking methodically and repeatedly about 4 inches to the right of it searching in the dark for the on/off button with my foot. Come to think of it, it's a pity the same isn't true for the video recorder. I think it would be very therapeutic to give it a good kicking every night before going to bed.

Video Conferencing

Two decades ago I used to watch 'Thunderbirds'; a science fiction puppet show for children. They had video conferencing. The picture on the wall of Scott had eyes that would suddenly light up and then the picture would become a video screen allowing him to video conference with his commander.

Two weeks ago, in the (rather small) meeting room of a client, I found myself taking part in that science fiction world of video conferencing. I watched them setting up the equipment as the 'real' people drifted in (new technology; the 'real people', present in the flesh, and the 'video people', present on the screen). Ten real people were expected and video people in Hong Kong and America. Part of the equipment was a solid tripod with a gun-turret like video camera in it. It was possible to remotely steer and zoom it and then to record the parameters as a preset. We spent half an hour setting up useful presets while the video person in Hong Kong was busy with his cables. Preset one was the chairperson in close-up, two was the presentation screen and so on. When we were done we skipped from one to another while the real people who were watching *oohed* and a*ahed* like spectators at a fireworks show.

After more setting up and contacting the parties in America all ten of us real people watched the pictures coming in from Hong Kong; they showed a very tired Chinese guy trying to sort his way through a spaghetti of cables so that he would be able to see us. Unlike Scotts' eyes his didn't light up, they just looked sunken and tired. While we were waiting we tried to impress the newcomers in the room by skipping through the presets on the remote-controlled video camera. For some reason most of them now seemed to include large amounts of the ceiling. Checking the camera revealed why; in an enthusiastic attempt to clear extra space in the room one of the attendees had shifted it out of the way and leant it against the wall in the corner!

Progress was slow, Hong Kong still couldn't see us and America was only present on audio. We decided to go ahead without video, and just use the phone links. However we rapidly became aware of the ragged breathing of all the remote attendees. 'Hello guys' shouted our chairperson, 'we can hear you breathing, can you press the mute buttons or something until you actually have a question to ask?' A chorus of 'yeah' came back followed by a wave of tinny, easy-listening music. If it wasn't enough to be spread across the world at ungodly hours, we now had to listen to the 'on-hold' music of one of the many telephone systems involved! 'Guys, someone's playing music. Can we stop that?' 'Why? Don't you like it?' came the reply.

The rest of the presentation went well, our chairperson mediated skillfully between the real people and the video (actually now audio) people. Eventually some of us suggested the idea of a break in the proceedings, and after a quick discussion we all piled out of the crowded room for coffee and a leg stretch. As we are wandering down the corridor someone shouts 'what about the others?' Of course; they wouldn't have heard the discussion, they would only know that it had all gone quiet at the other end. One of us ran back in to shout an explanation to them.

All in all it was a very educational experience, and not just for the content of the course being given. The problems seemed to stem from the fact that there were two groups of people involved, one of which was disadvantaged. One joking suggestion was that in order to make it all run more smoothly everyone should be equal and all the real people should go into adjacent rooms and also work through video link-ups!

In fact being present at such a mixed video conference as one of the video people must give you a very good idea of what it is like to be physically or mentally disadvantaged in ordinary social/interactive situations where the majority are not disadvantaged. You struggle to take part, you are often forgotten and you have to depend heavily on having a representative to assist you in the group of non-disadvantaged people.

The experience should be obligatory for anyone calling themselves a user interface designer.

Cafés

What is the essence of user interface design? What is the one project that is the quintessential 'design for use' problem? If you were to ask me that (and as yet no-one has ever done so) I would reply that it is the design of a good café.

This is a subject that I have spent many hours contemplating and discussing. Mainly because the only time I have free to contemplate and discuss is when I am sitting at a café. A good café with a terrace (sitting at tables outside) is a combination of a number of key elements. One of the main ones is the division of space into busy and relaxed areas. The area of the café and the terrace should have a definite feel of relaxedness and unhurriedness, while the adjoining area, the street, square whatever outside should be thronging with life, and by thronging with life I mean people, bicycles, shops etc. I don't mean cars and lorries. This division should be quite pronounced and sharp, business happening a couple of hundred meters away is wasted, the closer the better as long as the calm of the terrace is not affected.

The nature of the 'busyness' being observed is also important, people just walking past is not really too interesting. It is better to have a café and terrace facing some sort of 'happening zone' such as a junk market, a junctions of roads, a collection of local shops or even an area with paid parking where people aren't paying.

In non-ideal weather conditions coziness is king. In the early summer months I have seen terraces with heaters to take the edge off the chill and my favorite combination is to be on a covered terrace on a warm day during a summer rainstorm, just watching people scuttling about while I sip a hot cappuccino, I love it!

All the references to busyness and activity may lead you to think of cafés in the midst of city centers, but there must be a balance, the designed edges of the environment must be offset and softened by nature, some fine old trees, or some grass and shrubs. Even large potted plants make a difference. My theory is that the incredible detail and structure in nature is somehow visually restful. Far more so that a bland stretch of concrete. Your average tree has infinitely more detail than any man-made structure and the eye somehow knows this and can relax into it (I know I can spend hours staring at the natural forms of the flames in a good log fire but I get quickly bored when gazing at the central heating fittings). Furthermore with cafés that are visited regularly the presence of nature accentuates the awareness of the seasons which is also (for some unfathomable reason) important.

Personal comfort in a café is influenced by the environment and the more direct the contact with the environment the more of an influence it can have. The most direct part of the environment is the table and chairs at which you sit. Those trendy chairs with the uneven chair back look wonderful, until you try and hang your jacket over them, it seems to hang fine but as soon as your back is turned the jacket is somehow jettisoned onto the café floor.

Worse than trendy chairs are wobbly tables. If the wobble is not too severe and there are beer mats on hand then it can actually be positive, guests can level the wobble by jamming beer mats under the legs giving them a warm feeling of configuring their own environment. If however the table has all feet firmly on the floor and the wobble is in the table itself it can be terrible, as you shift and lean the table jerks from side to side, never actually spilling anything but completely ruining any feeling of calm built up by all the other factors. A plentiful supply of beer mats with one side blank is also a useful aid for technical discussions. Richard Bird once said that to be a good mathematician you must posses the ability to write upside-down on damp beer mats.

The feeling of being pampered is a more tenuous part of the overall atmosphere, this feeling stems directly from the staff that are waiting on you; their attitude and manner and the way in which your order is presented by them, the neat cup of cappuccino with its accompanying biscuit, sugar and spoon, delicious croissant warmed and served with a serviette, butter and jam. Evening liquors served with all the right accompaniments. Pure heaven! Finally, If anyone knows of grant-awarding bodies willing to sponsor further field research in this area I would be delighted to hear about them.

Food

What's the difference between a lamb and a salmon? In terms of the taste and the sort of culinary creation that can be produced with them there is a world of difference, but in terms of how the two are pronounced in the Dutch language there is very little difference (zalm and lam) and if you just happen to be pronouncing them in a busy and loud restaurant then the difference is almost imperceptible.

Those readers familiar with working in restaurants will already be wincing at this point. The danger is that the waiter will stick his head through the kitchen door and yell 'one steak, one lamb'. Twenty minutes later he'll walk to the customer's table with one steak and one salmon!

What's the solution? Start running a vegetarian restaurant? Shout louder? No. The answer is to redesign the language. In Dutch restaurants if there is a lamb dish on the menu it is often given a different name, the name of the dish, the French name for lamb whatever, as long as it isn't called lamb.

When you get down to it this is effectively the design of syntax and semantics of a dedicated command language (but don't try to tell this to the waiter). It is the configuration of a system of communication to make it more friendly to those using it. The nature of the problem is that you have two very different dishes with very similar names. A small mistake in yelling the name or in hearing it can lead to a big difference in the results. The syntax elements are similar but the semantics are very different.

Similar things happen with information systems. That terrible program where the bottom choice on the menu is 'save all information' and the one just above it is 'exit without saving'. The commands are syntactically close (a matter of shifting the mouse 2 millimeters), but they are very different in the results, a combination that leads to disastrous results.

Peter 'hips' Boersma used to work in a pizzeria and there they developed a pizza shorthand. The idea was that each pizza had a name that was totally unconfusable. This was not as easy as it sounds. The 'quatro stagioni' was obviously shortened to a 'quatro', and to avoid confusion the 'quatro formaggio' was a 'cheese'. For a while the 'quatro salumi' (something to do with sausages I think) was shortened to 'salumi' but this was syntactically too close to 'salami'.

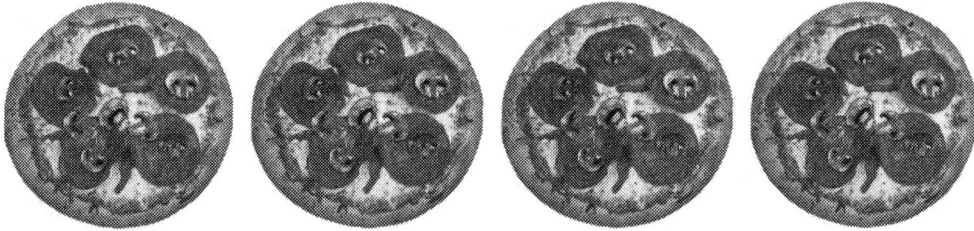

Another pizzeria trick was the garlic. Customers could choose pizzas with or without garlic sauce. The garlic sauce was invisible and if you had two salami pizzas, one with garlic and the other without there was no way of telling which was which without starting to eat them. A golden rule for feedback for software, products or even food is 'make the invisible visible'. Visible information is faster to access and process. The solution to the garlic problem was to put visible bits of garlic on those pizzas treated with the invisible garlic sauce. In fact, when it comes to cooking, feedback plays a vital role in all its forms. It is one of the few areas where taste feedback is used in controlling actions; seasoning a dish and tasting it until it is exactly what you want. Tactile feedback is also used; stirring a sauce until you feel it thickening against the spoon. Audio; turning the gas up or down purely on the basis of the whistling noise that the gas jets are making. Smell; something's burning. And visual; basically watching what your doing!

All this feedback is somehow natural. In contrast the world of software and products uses feedback that is artificially added to the system in order to help the user: the beep signals, the red LEDs and so on. However, there are instances in the culinary world where artificial feedback is added to a system in order to help the user. Welcome to the 'white sauce syndrome'. The chef is cooking a set menu, she has a white sauce for the fish, a white sauce for the cauliflower and a white custard for the dessert. If you are going to pour one of the sauces on the fish you need to know which sauce is which, obviously they taste different but to make things faster you need some visual feedback. Thus cooks will often modify the recipe so that they can see what is going on; putting a few bits of mint in the custard or capers in the fish sauce.

So the next time you're in a restaurant spare a thought for the poor cook struggling to meet your order in a steamy, noisy world of poor feedback and badly designed languages!

Infra-Red

I used to baby sit for a family that had one of the latest models of television, it was in color (this is going back a bit) and it had a remote control. This was a very early remote and it communicated with the television using bursts of high-frequency sound, the protocol was very simple and supported only very simple actions like click once to switch to the next channel and click twice to adjust the volume.

These days, short-range communication between gadgets is carried out with wireless networks and wireless Bluetooth communication.

In between the two there was a long period where infra-red light was used as the communication medium. It was easy to produce, easy to detect and wasn't perceivable by humans and so it wouldn't distract anybody when in use. In fact sometimes it was so imperceptible that people didn't even know it was there. Commuting on a train once my table was shared by two people – colleagues of each other – who immediately whipped out their laptops and started ticking away. 'Oh!', Said one of them, 'I've got a strange folder on my desktop.' 'That's strange', said the other, 'so have I'. Clicking away at their respective new folders eventually led to the realization that their two laptops, back to back with one another, had struck up a communication via their infra-red ports with the result that they both now had shared folders on the desktop. The imperceptible nature of infra-red light coupled with its remote nature meant that as it became more widespread there were more opportunities to abuse it. You didn't have to be physically connected to communicate, as exemplified by the rather antisocial behavior of a young, newspaper-delivery boy who did his paper round with his remote control device in his pocket. After delivering the newspaper he would peep in to the lounge window and if the TV was visible he would try zapping it with his remote control, turn it on, fish about for a channel and turn the volume up full. Victims would never realize what had happened and would probably put it down to a strange technical glitch with the TV, as the idea that someone can interfere with your property without actually having physical access to your space is quite alien.

Even the car manufacturers got into the act and built infra-red communications into the key fobs so that you could turn the car's alarm on and off and even lock and unlock the car as you walked away or approached your vehicle. A system that was very easy to use and thanks to the encryption used in the signal a secure solution as well. At about the same time electronics manufacturers started producing universal infra-red remote controls. A nifty idea, the device would detect infra-red signals and store them so that you could play them back when you wanted to. In effect you could use them to record the infra-red codes from all your remote controls and then have just one remote for everything. Very nice indeed.

Mix the two previous devices together and you had the perfect scam for stealing cars; someone parks a nice car outside the theater and you sidle up to them with your universal, infra-red remote control. When they zap the car with the encoded signal to operate the alarm, you record the signal, wait till they have gone then replay the signal to switch the alarm off, open the car and away you go.

The message is that whatever communication medium is used it will have its own unique properties which can lead to serendipitous advantages or sinister disadvantages as different devices are developed that utilize the medium. The only real solution is good encryption of the data being communicated.

Strangely enough, even the old ultrasound, remote control had unique advantages that compensated for the simplicity of the communication. The main one was that the high-pitched sound used by the device was very similar to the high-pitched sound used by bats when they are hunting for food. If you were sat watching some boring program on television and there was a moth in the room you could use the remote to zap the moth with bursts of high pitch sound. The poor moth would think it was being targeted by an incoming bat and would immediately take evasive maneuvers. Television was never this much fun!

P E O P L E

One side of interaction are the devices that we have dealt with in the preceding part. The other side are the people that interact with those devices, and if devices are complex, then people are even more so. A true picture of the human user is a huge exercise taking into account a range of disciplines from the physiology of our senses through to emotional and sociological studies of user behavior.

Here we won't be going to such depths. What we will do is split our consideration up into two areas. Firstly, the wider area of people in their context. People surround themselves with things, the longer we stay in one house or office the more we acquire all sorts of things. Things to make our lives easier, things to entertain us, things that are beautiful, things that just haven't been thrown away yet. Not only are we surrounded by things, but we also exist within different groups of people: the family, the work group, the community. These columns deal with contact between people, products used by couples and the ever present problems of household junk.

The second group of columns here deal with subjects more closely related to how people do things irrespective of the context they are in. These include a look at how our senses are attuned to stereo vision and sound and a consideration of children and interaction – a very rich source of interesting observations as children's reactions are less influenced by previous experience simply because they have had less. In particular we shall have a look at children and the issue of pointing.

Contact

I've just been watching the movie 'The Graduate', for about the seventh time. Highly enjoyable and thought provoking, and this time there was a thought provoked about media and communication. In the movie, the parents of the romantic young man and woman have their own plans for them and decide that they should not see each other any more or have any other sort of contact. In that era it would not be too difficult, it would just be a question of turning them away if they called at the door, hanging up on them if they rang and intercepting letters brought by the mail man. There were few 'gates' of communication and the parents were gate-keeper to them all.

Nowadays that has changed. The increasing spending power of average Western individuals coupled with the decreasing costs of technology means that parents are no longer the gate-keepers of the communication channels. Indeed, all to often, the younger generation have more access than the parents. ('Hey! Mom. I'm sending an email to cousin Jenny, is there anything you want to say to her Mom?'). Furthermore, the technological gates themselves are more discrete and more portable. It used to be the front door and the fixed point telephone. Now it can be the mobile phone in the bedroom, the SMS message in the garden, email at the internet cafe, ICQ at work.

As well as more channels of contact, the new technologies have meant that it is easy to maintain contact after a chance or casual meeting with someone. Exchanging telephone numbers is very difficult without a pen and a piece of paper. Occasionally, if I'm with a group of people I've tried

the approach of allotting two figures per person, but it never seems to work. However, email addresses can be easily memorized, and if you know where the person works and what their full name is you can even take a fairly good guess at the email address.

Nowadays, new technology is often responsible for that chance meeting in the first place. You can search for old friends on the internet, in chat groups you can chat to people with similar interests (think of the movie 'You've Got Mail'), ICQ offers a random chat facility and there is a new WAP 'blind date' service being launched in the UK where you can enter a personal profile, and as your WAP phone is aware of where you are, you can be put in touch with someone with a matching profile in the same vicinity.

This idea has been around for a while but only now does it seem to be taking off, mainly because it does not require buying special dedicated technology. Other contenders have been the Japanese 'love-getty' a rather less discerning version that just used to beep when you got within 10 meters of someone else who had one of the things. The consumer electronics company Philips also had the visionary idea of 'hot badges'. Somewhere in between the two mentioned above, hot badges were badges that could be profiled and would alert you if you came in the vicinity of someone with a matching profile.

To some extent this opening up of communication gateways and their adoption by the younger generation is a revolution comparable with the role of the car in the fifties. All of a sudden teenagers then had a private space that was their own, outside the scope of their parents' space and where they could meet other boys and girls. The sort of set up that was typified in the movie 'American Graffiti'. This has meant that today there is a certain generation that is used to hearing their parents saying, 'Oh yeah, your father and I used to drive down to the lake together in his '56 Chevy.' Maybe with the current contact revolution there will soon be a generation that hears their parents saying, 'Oh yeah, your father and I used to meet up every week on the z-world chat-zone after we'd connected our 28k modems.'

One final note on the subject of romantic communication. Everyone has an old box of love letters, and you can certainly have romantic phone calls, you can even send romantic emails (as long as they don't get forwarded around the globe!) and you can be a slave to romantic SMS messages, but has anyone ever carried out a romance using the good old fax machine? I wonder.

A User Group of Two

There are certain things that are difficult to define: freedom, beauty, truth... Well, here's a new one: toast. Toast truly is in the eye of the beholder, ask any couple living together. I like toast to be real toast, Wendelynne however enjoys slightly warmed bread for breakfast, and accuses me of burning the toast every time I make it. Although not as fundamental an issue as freedom, beauty or truth, I'm fairly confident that more time has been spent discussing (or arguing) about this issue than any of the others. However with the advent of clever technology we can now side-step this and discuss things like freedom, truth and beauty at breakfast over perfect pieces of toast.

Much of today's new technology is intimate, it is carried, worn or intended for personal use by one user. Think of mobile phones, laptops, organizers, MP3 players. However, there is still a lot out there that is less intimate, it just sits around in the kitchen or bedroom until someone, anyone, comes along to use it, and although it is designed for one user, it usually ends up in domestic environments inhabited by a couple. This needn't be a problem, unless the technology needs to be used independently by both people at the same time, like the bathroom ('Well what are you doing in there?'), the phone ('Shall I tell you how long you were on for?') or the television ('No I don't want to watch it, it's just monkeys and space and nothing happens in it').

More general problems do arise with technology owned by couples if that technology can be parameterized - which is jargon for setting-up various things before you actually use the technology. Think of setting the dial on the toaster for how well-done the toast will be. ('Hang on honey, I'm just parametrizing the toaster!'). This is when you realize that you both have different ideas about what it should be set to; that you prefer real toast and your partner doesn't. Or you realize when you share the same push-bike or car how short your partners legs actually are, or that they prefer showers that are scalding hot and leave the shower set on 'scalding hot' when they have finished, that they must have incredible hearing because they only need the TV volume on 2 which makes you think it's bust when you come to switch it on, and that they need so many lights on to read that the lounge looks like something from 'Close Encounters of the Third Kind'. And what about different wake-up alarm times? There is only one thing worse than getting up at 5:30 in the morning and that is getting up at 5:30 and then using half a sleep-starved brain in total darkness to try and reset the alarm for two hours later for your partner who is blissfully under the covers making snugly 'I'm enjoying being asleep' noises.

How can designers solve the problem? Well, there are the occasional un-designed solutions to these problems. Julia and David had their TV stolen but the thief couldn't find the remote control (let's face it; who can?). With the insurance pay-out they got the same model and so ended up with two remote controls, one each, which may seem like the ideal way to solve the issue of who has the remote, but in actual fact turned out to be quite unworkable! However, designers in the commercial world are coming to terms with this dual-ownership problem. You can now get a 'his and hers' toaster with two presets, ensuring that one partner can enjoy limp, warm bread while the other tries to spread butter on brittle carbon. Also, alarm-clocks are available with two presets for the alarm and (expensive) cars have a number of different preset memories for the seat configuration. Any product designers wishing to make the next breakthrough in this field need only study a couple living together and identify all the areas where there is some disagreement in how the technology is to be used. Even simple, non-digital technology is open to redesign. You can't get more non-digital than a duvet, but there is a company that manufactures duvets that has finally tuned in to users' needs and started manufacturing a double-duvet that is thick on one side of the bed and yet thinner on the other; brilliant!

The really interesting thing that is probably going to happen in this area will be the Bluetooth revolution; we'll carry round our preferences for everything in our mobile phones and they will be instantly transmitted to any bit of technology we use; we'll never get a badly cooked piece of toast again. The corollary to this is that parameterizing your mobile phone will probably take about three weeks: 'What time should the alarm ring in the morning? What TV channel do you like watching, and at what volume?, Do you prefer milk in your coffee? How do you like your toast...' And just imagine what will happen then when your partner needs to borrow your mobile!

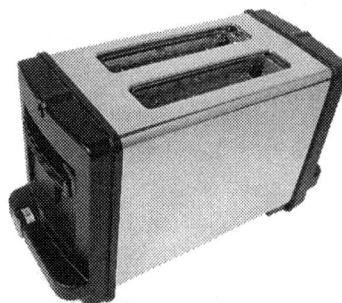

Junk

I have a lot of stuff in boxes. I also have some stuff out of boxes as well. All the boxes with stuff in are in the box room along with a few old computers one of which has the first version of this article on its hard disk, so I shall have to start again. Fortunately I have an ancient laptop as well as the two ancient computers in the box room and what better subject to write about than all those boxes.

Part of being human and living this Western civilized life is the accumulation of things. American families often store excess junk in large lock-up units, if they default on the rent then the contents are auctioned off. Punters view the contents through the open door without actually entering and bid on what they can see. Strange stories abound like the one lot that contained only three boxes. One of the bidders thought he could make out a couple of rifles in the far corner which could mean the boxes contained the uncollected spoils of a robbery. He bid and won the lot and as he walked in his feet started slipping and sliding; the floor was strewn with mothballs. Only when the cases were opened did it all fall into place. The crime wasn't robbery but murder and the mothballs were to mask the smell of the three bodies in the cases.

Well I certainly haven't got any bodies anywhere (although I haven't seen the cat since we moved) and I don't have any mothballs or rifles. So what do I have then? Well I try (rather unsuccessfully) to follow the old adage about only having things in your house that you know to be useful or believe to be beautiful. However this seems to ignore things that will probably be useful in the future, and things that are easier to store than to have collected or trashed. Another important omission is things that have a sentimental value. I have a rivet from the Eiffel Tower somewhere, it is caked in rust and certainly not beautiful. Furthermore its usefulness as a rivet is pretty limited, the reason probably for its being removed from the Eiffel tower in the first place. However I am not about to part with it, I think it's great.

Occasionally I find parallels between all this real junk and the virtual junk on my computer. My hard disk is full of directories called: old-project, old-old-project, project-mar-97, misc., sort, check, trash and so on. This is just about manageable if I am dealing with written projects with small text files, but it runs into problems if I have to deal with multiple copies of video and sound files. Indeed whereas text files take a lot of work to create them (don't laugh it's true!) huge megabyte amounts of video and sound can be recorded just at the touch of a button. Sometimes I need to tidy it up.

The question is, can we look at the situation in the real world and see if any of the solutions translate to the digital world? One real world problem is the putting away problem, not taking the trouble to archive stuff because it will probably be useful in the near future; this leads to clutter build up on your (physical) desk top and in your office. My solution is big tidying binges (usually as an excuse for not getting on with writing) where everything on my desk top usually ends up getting stuffed in a big file labeled 'desk top March 2000'. This works in the digital world as well.

There are two other solutions which do not involve actual tiding up but instead render the junk more comprehensible on some level. One is piling things up; you associate stuff and gather it into chunks, these are conceptually easy to deal with. So instead of a sea of documents you have just eight piles with easy conceptual labels. The other solution requires even less work; you familiarize yourself with the junk. If I spend two hours sorting through a cupboard sometimes I can end up throwing hardly anything out and yet by having considering everything, the contents no longer feel like anonymous junk but a known collection which is less daunting.

Conceptually speaking physical tidying up always seems easier than digital tidying up. It is easy to adopt a one-pass process when you are dealing with a simple spatial model; first upstairs then downstairs, first this room then that room and so on. Even the checking is easy; have I done this room already? Take a look! The visual difference between a tidy room and a not yet tidy room is huge. Although most computers now offer visual / spatial ways of dealing with files the process of stepping through a collection of nested directories with long-forgotten names is far more difficult.

The ultimate solution in the real world is just to move to a bigger house. This translates very easily to the digital world, simply do as I did, buy a new laptop and stash the old computers in the box room... but remember to copy any documents that you are in the middle of writing.

Marks and Scratches

I am of the generation that has a 35 millimeter SLR camera instead of a compact 35, an APS camera or even a digital camera. The nice thing about the old SLRs was that they were robust, lasted for years and as a result quickly got covered in marks and scratches. Indeed, a long time ago the camera company Pentax ran a series of adverts showing the Pentaxes of the stars such as 'Spike Milligan's Pentax, Alan Whicker's Pentax' and in the picture a Pentax SLR camera appropriately beaten up, marked and scratched by years of reliable service with the renowned owner.

These marks and scratches are a form of personalization, not explicit as in 'I want this mark here and two of those long scratches here' (although there could be a market there!). These marks are what distinguishes your mass produced SLR camera from the millions of others of the same model.

It is interesting that you don't have the same movement when it comes to cars, usually a dent or scratch is a reason to get the car to the garage as fast as possible to have it repaired. Why are cameras different? I reckon it has something to do with creativity.

What about archeologists and trowels then? They have a big thing about trowels, there is nothing worse than turning up at a dig with a shiny new trowel. Apocryphal stories abound of archeological volunteers using angle grinders and mud to give their new trowels the same aura as that of the dig leader. Are they creative? Or has it more to do with experience? Using a camera is a solitary occupation, only occasionally do you go out with anyone else toting a camera, whereas on a dig you are surrounded by crowds of other diggers all eyeing up each others trowels.

The industrial/interface designers associated with that gem of organization the Psion went in for personalization. They go by the name 'therefore' design, which sounds a bit strange but it does meant that they can use the mathematical three dot therefore symbol as a trendy logo. Anyway, personalization was part of their design, not just in the software interface but also in the hardware interface, in an implicit way, the casing was coated in a rubbery plastic substance that was designed to wear down in a personalized sort of way.

As well as this rough and tumble personalization there are many instances of more explicit personalization.

Personalization of our living space has always been a trend, with the current spate of television programs covering house and garden makeovers people are saying that gardening and interior design is the new rock-and-roll of our era.

The technological world has finally tuned into this human desire to personalize one's environment. Take the mobile phone market for example. After the initial wave of mobile phones people started customizing them with leather carrying pouches, until the phone manufacturers got into the act with clip-on colorful phone shells. Even Psion joined in with colorful clip-on plates for its Sienna range.

Then there are ringtones. Ten years ago there were the first doorbells that you could configure to play the tune you wanted. Now seemingly everyone has a mobile phone that plays a different piece of beepy music instead of ringing, this is good design in that it aids recognition of whose phone is ringing, but there is also something nice and cozy about setting up your telephone to ring in the way that you want it to ring.

The flexibility of the computer meant that the on screen world was an ideal arena for explicit personalization of working environments. Just look at the difference in personalizing the Mac software environment and personalizing the PC environment.

The early days of configurable color windowing systems saw a boom in the effort people put into selecting their color coordinated background, borders etc. This movement reached its peak in the research world with the WOOL object-oriented graphical window system where every last pixel could be customized. It probably set computing research back by ten years due to lost research time as the entire research community was defining background tiling patterns, frames with elaborate borders and detailed corner pieces. It was far worse than any virus.

Take-Out Service

If you are providing a service to a user and that service is embodied in a public piece of hardware how do you know that the user won't invent their own service and just walk off with the piece of hardware? The short answer is that you can never be sure. So how can you prevent it in the design of the piece of hardware? Well, there are a number of solutions:

1) Don't offer the service in the first place. No one is going to walk off with your tea-spoons if you haven't got any to begin with.

2) Charge a deposit. Two dollars to use a luggage trolley means that people will be less inclined to use them for playing bumper cars in. The problem is that at that price people are going to be less inclined to use them for anything, even carrying luggage in.

3) Trust them. Not everyone is going to be a cad and take your pens, only a small few.

4) Make it a part of the service. Relish in the fact that people like your cups and saucers so much that they will want them for their own. Plaster them with adverts so that you can at least get some publicity out of it all.

5) Make them not worth taking. If the cups and saucers in question are chipped and cracked and greasy who in their right mind would consider misappropriating them? After a month you would still have all your crockery but would have lost most of your customers.

6) Nail them down. (The hardware, not the customers!) Put the pens on chains, put the tea-spoon on a chain, attach the computers and keyboards to the desk with super glue.

7) Make them disposable. If they are plastic and cheap it doesn't matter what happens to them. Only trouble is that things that are plastic and cheap seem ... well ... plastic and cheap.

8) Make them unusable out of context. Aha! This is interesting. Design them in such a way that they function correctly in the context they are offered in, but that they lose the functionality if someone takes them home. Classic examples are the shoe shop with just one example of each shoe on show. Steal one to take home and it becomes worthless. Although the possibilities for a one-legged kleptomaniac are astounding.

Sleeper trains always try and offer a luxurious service, but to stop the passengers making off with the lovely, heavy, china cups they come equipped with a hole in the bottom and a thin disposable plastic liner. You are drinking out of a piece of quality earthenware, but try taking it home and it becomes useless. The only people even remotely interested in stealing such things are usability people like me who are fascinated by the implications of the design. In Russia I have heard that they used to operate a similar system with soup spoons in the diners. They would be drilled with a small hole to render them useless as soup spoons. To actually use them the customer would plug the hole with a blob of bread. The plug would just hold long enough to get through a bowl of soup before disintegrating.

Clothes hangers in communal wardrobes have to be removable by the very nature of their function. Fit them with the usual hooks and they can go walkabout. Fit them with an unconventional hooking method and no one will bother taking them home.

Dedicated headphones can be made with two, fixed, mono jacks side by side instead of a single stereo jack. You can't take them home and plug them into your CD Walkman, and if you do decide to just use one channel and plug one of the jacks into a mono socket you are prevented from doing even that by the fact that the two jacks are fixed side by side. You can't plug one of them in because the other will poke up against the plastic next to the socket you are using.

Isn't it funny what goes through your mind when they give out the headphones for the in-flight movie?

Stereo Vision

How many of you have heard of the book 'Flatland'? Written in Victorian Britain by Edwin A. Abbott, it tells the story of a two-dimensional world peopled by shapes. The higher nobility are shapes with lots of sides, the upper class are pentagons, the middle classes are squares (real Victorian stuff this!) and the lower classes are triangles, the lower they are the thinner the triangle, criminals are incredibly thin triangles, and then the women are straight lines. A wonderful exercise in two-dimensional topology but a truly sad reflection on the values of those times.

Thankfully we live in a 3D world (and a world where women have a somewhat better perception in society). But is it really all that 3D? How often do we go swooping around in the third dimension? When was the last time you did a 'loop the loop'? We actually live a sort of 2½D existence. Confined to live in a thin, 3D layer occasionally going upstairs to another thin, 3D layer that just happens to be located above the first.

You can tell we have always been confined to this 2½D existence in the way our ears have evolved. We can gauge direction with our ears thanks to some complex integration between the signals reaching both ears. But our ears are separated on the horizontal plain meaning that we are sensitive to direction on the horizontal plane but on the vertical plain there is no vertical separation between our ears and thus we cannot pinpoint vertical direction.

In practical terms what this means is that we get confused when someone shouts to us from a balcony. We can tell that the shout is coming from a particular direction but not from how high above the ground it is coming. One solution to this is to separate out ears in the vertical plane. This sounds painful but it can be done by the simple act of tilting our heads to one side. Cats and dogs sometimes tilt their heads when listening so there could be something in it.

What about eyes? Do they triangulate in the same way? Are our eyes geared up to life in the plane? Well, due to the difference in velocity between sound and light

they can't do the same trick that the ears do, and anyway, they don't need to. Direction in vision is implicit, you don't see something and wonder, 'Which direction is that in?' You can see which direction it is in!

If we can judge direction with just one eye then why do we have two? One factor is redundancy. One only has to think of Jason and the Cyclops to realize the drawbacks of monocular vision. But binocular vision is useful for other things. We get different views from different eyes and the integration of these two views provides us not with direction information, but with depth information about what we can see.

That is the key to why our two eyes are next to each other and not one above the other. We need to see what things are in front of other things and that depends on how the edges of things overlap each other. When things are in front of one another they have more edge overlaps in the vertical plane than in the horizontal plane. In general things are tall and thin and not short and fat. Think of tree trunks, cliff walls, people, edges of mammoths etc. Furthermore, those things that move are usually, like us, confined to moving on the flat, they don't go higher or lower but they do go from side to side. And we ourselves are going to navigate in this plane going in between these things we can see.

This integration of views from two eyes is just one of the factors that govern our perception of depth, there are many others. There are cues that are inherent in the view itself. The loss of detail and the blue shift of color as things get further away. The brain is also aware of how much our eyes have to move together when we look at something. This sense can get confused when we look at patterns that repeat horizontally like curly telephone cords, mesh fencing or random dot stereograms, making them look nearer or further away than they really are. Another small detail that the brain is aware of is the amount that the eye lens has to stretch to bring something into focus. Ever seen orange icons on a blue background? The orange things really do seem to float in front of the blue even though they are the same distance away. The difference in wavelength between blue and orange light means that the eye lens has to do different amounts of stretching to focus them and thus the brain is tricked into perceiving them as being at different depths.

In conclusion then we have evolved in amazing ways over millions of years to integrate all sorts of information regarding the world around us. We have evolved to be almost perfect. All we have to do now is sort out the last vestiges of that Victorian attitude to women and we will be there!

Being Overheard

The ability to record and transmit speech from one place to another has a very checkered history. From its humble beginnings with the phonograph and the telephone we have now reached a stage where our lives are awash with things that transmit or record speech. The key issue is not so much: How much our voices are being monitored? But rather: Are we given feedback so that we know when we are being monitored? Call up any telephone help-desk and the chances are that the phone call will start with the phrase; 'Some calls may be recorded for training purposes'. Once this has been stated it can quickly be forgotten. Contrast this with holding a face to face conversation while someone else is in the room listening in. Their continual presence reminds us - visually - that there is someone hearing our conversation.

This column was promoted by the current court case against a company called Marvel. The company expressed an interest in buying out Jasmine, one of its rivals. Contact between the two companies was very cautious as Jasmine didn't want Marvel just finding out about all its good ideas, copying them and then withdrawing from the takeover. Well, on one occasion three execs at Marvel were in their conference room trying to contact the legal chief at Jasmine to arrange a meeting with Jasmine's engineers. They called her on the speaker phone and as she was out of her office they left a message on her voice-mail requesting the meeting. When they had finished they got chatting amongst themselves and started discussing their real plans which were indeed to try and get hold of Jasmine's ideas and not do the deal on the takeover. They had it all worked out. They had overlooked only one small detail – they had forgotten to switch off the speaker phone which was still live and feeding their conversation into the voice-mail of Jasmine's legal chief. Needles to say when she listened to her messages the next day all hell broke loose!

The act of being overheard without realizing that you are being overheard has been used in many movies and soap operas. From the couple having a private conversation in the babies bedroom and not remembering that the baby monitor is transmitting it downstairs, through to the cop movie where the villain's private admission of guilt is picked up on the mobile phone that the hero has turned on in his pocket. In well set up recording studios they always have a big red light that is on when the sound is being broadcast. This is a universal signal that is used in other systems, including some phone systems and many on-screen recording utilities. When things are being recorded or broadcast out 'in the field' such luxuries are often not present and there are notorious examples of leaders making off the cuff remarks after speeches not realizing that the microphones are still live. Both George W. Bush and John Kerry have been caught out with this in recent years!

I have also seen a copy of a classic video, made accidentally by two old ladies who borrowed a video camera to send a tape of themselves to a relative. They set it up on the tripod and then tried to figure out the remote control. Within a few seconds they had managed to turn the thing on without realizing it and it then proceeded to record their in-depth discussions about which button to press to make it work. Wonderful stuff. But, without a doubt my favorite story on the theme of 'listening in' was from the Moskva Hotel in Russia. During the cold war it was the place where foreign reporters were obliged to stay. Needless to say they all assumed it was heavily bugged. One journalist was determined to find the bug in his room and eventually discovered a suspicious looking metal plate under the carpet. He had a screwdriver with him and used it to unscrew the plate, and after a struggle was rewarded by a distant crash as the chandelier in the state room below crashed to the ground.

As audio and video systems become more widespread and more well integrated into existing technologies we are moving toward a point where everyone can be observed all the time. Certain organizations are already cracking down on the use of camera-mobiles on the premises and in the UK the secret service has banned staff from bringing Furbies into the building.

In such situations interaction designers have to keep an eye on two things; firstly they must be very careful when designing such technology to think about letting people know that they are being observed, but more importantly they should also become involved in the widening debate about the ramifications of such technology in our everyday lives. It is true that closed-circuit TVs are good at supplying video footage to trap thieves, but when a stranger with a mobile phone can snap off a few shots of you at the local gym it is a different matter altogether.

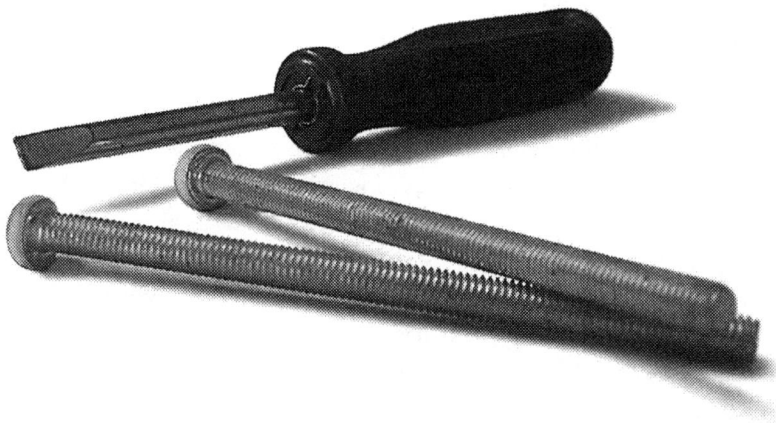

Children

It is late in the evening. I am siting on the sofa in my sitting room with a lovely redhead. We have just finished a delicious, candlelit meal. She looks at me with come-to-bed eyes and I open my mouth and say 'shall we go up the wooden hill to the land of beddly bobos?' What on earth is going on? Well, it's called convergence. The woman in question is my partner and the strange language I am uttering is baby-talk, and I'm still uttering it even though the baby was in bed 5 hours ago. Readers with young children will be intimately familiar with the effect and user interface designers with toddlers will also know what a wealth of interesting observations they can give rise to.

Back to convergence; it is the linguistic effect where party A (Morgan) starts picking up the language used by party B (her parents) and vice-versa party B starts using terms from the vocabulary of party A. The end result is that party A starts improving her language at an alarming rate while the language of party B degenerates into a sort of pidgin English mixed with silly baby-talk and animal noises. Other examples are 'cockanellie' which we use because it sounds far nicer than cockerel and 'toothy-peggies'. However, my two favorites are 'tell 'em phone' for telephone, and 'biayer' - a word whose meaning is uncertain but, judging by the frequency with which it is used, is vital in day-to-day communication. In fact I sometimes wonder how we grown-ups can possibly communicate without it.

A user interface designer experiences pleasure in the strangest of things: going up and down in elevators, copying just one sheet of double-sided paper on a complex copier, getting lost in airports; and they will positively drool to try out the latest drinks vending machine. A new addition to my list is the fascination when, in the middle of a television program, the baby begins to adjust the volume and brightness controls. The question of what Sam will do with the passage to America is immediately dwarfed by the question of whether the baby realizes the results of her actions? Has she a preference for bright or dull pictures? Does she turn the controls both ways? And she also changes the channels. What channels does she prefer? A brief, and totally unscientific study yielded the surprising result that she liked watching rugby; a game similar to American football but with less padding and less rules.

One can learn a lot from a baby performing even an act as simple as screwing tops on and off (in the approximate ratio 10% to 90%). She has a definite preference for using hands in one direction only, in other words if she wants to turn something clockwise she uses her right hand

and if she wants to turn it anti-clockwise she uses her left hand. Could this have anything to do with the origins of our conventions for clockwise screw up / anti-clockwise unscrew?

Living with a baby makes you more aware of the detail of your surroundings. Morgan continually notices minute details that grown-ups tend to skip. One evening she suddenly began shrieking 'teddy' at the top of her voice in a restaurant, it was only after a detailed reconnaissance of the surroundings that we discovered one of the guests was wearing a jumper with a very small teddy motif on it. This awareness of detail extends to other senses as well. With grown-ups, sound seems to be easier to screen out than visual information, but with a baby both seem to be on the same level. When Morgan hears a dog she will say 'doggy' in the same way as when she sees one. Often we will be looking round for a doggy before tuning in to the audio channel and realizing that there is a dog barking in the background. This importance of sound in the babies' world is further exemplified by the sound-naming that babies use for things. A choo-choo train, a bow-wow, a baa-lamb, a moo-cow.

Knobs, buttons and noises are pretty low-level stuff, a higher level example was the use of metaphors, bumping into a field full of Aberdeen Reds, Morgan referred to them as 'mummy cow, daddy cow and Morgan cow', using our family as a pretty direct metaphor for describing the cow family.

Well that's it for now, all that remains is for me to say; bye-bye, see ya, daddy gone!

Pointing

A while back our usually comatose cat suddenly started chasing a piece of Lego about the floor. 'Look at the cat,' I shouted to 10 month old Keiran and pointed at the cuffing animal. Keiran responded with a leisurely gaze at my outstretched hand and didn't notice the cat at all.

Now at 13 months he is starting to get the hang of pointing, and his struggle has made me realize what a complex activity pointing actually is. He has also started pointing at things himself, mainly because I immediately react by naming the thing that he is pointing to. All of a sudden he can interrogate the world of objects he is living in. Sometimes when he is outside in the pram he will point with both hands at the same time for emphasis, 'power pointing'. This whole pointing and interrogation idea reminds me of my own first contact with 'Balloon help' on the Macintosh. The freedom of being able to lazily swish the pointer around the screen and have the system declare what things were.

Coupled with Keiran's pointing is a new found interest in pushing buttons. (There is also a recent habit of trying to jam things that aren't videos into the video recorder but that's not particularly relevant here). There seems to be some connection between pointing and pushing buttons. For starters the act of pointing and pressing buttons demands the same action with the hand, but then so does scratching your ear, so it's probably not very significant. However there are functional parallels, pointing is indicating something to a third party so that the party can take action in some way, when Keiran points the third party is me naming something, when I point in the baker's, the third party is the baker getting bread rolls. Pressing buttons is the same thing but the third party is not the parent or the baker. Instead it is the technology with which the button is associated, and the user is not implicitly saying 'explain this thing' or 'get this thing' but 'carry out the action associated with this thing'. The fact that the user has to physically press something is only a restriction in the technology, sometime in the future I am certain that interaction with lights and other household systems will just involve pointing to them from a distance. Needless to say young children will have a field day with such systems!

Our other child Morgan (4) is also learning about pointing, but not pointing in the real world, but about virtual pointing with the computer and the mouse (a term she finds very amusing). When she asks if she can play with the computer my question of 'Do you want to draw pictures or spell words?' is often answered with 'No, I just want to click on things.' She actually enjoys just playing around with the windows based operating system. It is a good indication of the 'naturalness' of current computer interfaces that Morgan's learning of virtual pointing was accomplished in minutes, what she now enjoys doing is practicing and using the skill. In fact the only glitch in learning was up and down confusions while talking about the actions themselves mainly because of the separation of tool (mouse) and feedback (on screen cursor) and the fact that the 2D horizontal motion of the mouse is mapped onto the 2D vertical movement of the cursor. Left and right are no problem but telling Morgan to move the mouse up did indeed result in her lifting the mouse from the mouse mat. I was mixing the tool and feedback co-ordinates, 'move the cursor up' or 'move the mouse away from you' would have been better.

Funnily enough, they are not just busy with pointing in the two different worlds but they are also involved in other activities. Keiran is now staggering about the real environment picking things up and putting them down in different places, and last week I silently peeped over Morgan's shoulder to see what her 'just clicking on things' was all about. She had opened some random directory and was fastidiously moving all the icons one by one from the left side of the window to the right side. Did this mean that the organizational chaos wrought by the kids in the real world would soon be matched by organizational chaos in my digital world? Let's hope that I can at least write an article about the ensuing problems.

Left or Right?

When John was studying medicine, his group had a headless cadaver to dissect in anatomy lessons. John used to spend hours cutting and isolating skin and organs while listening to the lecturer; 'notice the organs relationship to the right lung…' For months he was told what to do on the left and right sides of this body, but the left and right of this body were the exact opposite of his own left and right. Eventually he lost track and would get confused about left and right when he was back in the 'real world'.

Surgeons must find ways of switching between the two frames of reference; that of self and patient. This issue of self or patient as the reference point is echoed by the situation where a group of people are talking about something big surrounding them such as a ship or an airplane. Just as there could be mismatches between the self and patient frames of reference so too there can be mismatches between the self and vehicle frames of reference. This is one of the reasons that some vehicles use the terms 'port' and 'starboard'. There was one airplane crash where this convention broke down; there was engine trouble, the co-pilot turned in his seat to try and see the engines through the windows and told the pilot which one it was. Instead of using port and starboard he used left and right and because he was twisted round so that his left and right were opposite to those of the plane there was confusion and the pilot ended up shutting down the good engine instead of the bad engine.

This confusion between self and vehicle frames of reference also comes about with coach tour guides; they stand at the front of the coach facing the passengers and say 'to the left is the Cathedral'. Experienced tour guides know how to clarify the problem and tie their commentary into the passengers frame of reference; 'on *your* left is the Cathedral'.

A more grisly example of the left and right confusion with bodies facing you was the account of an operation in a Dutch hospital several years ago where a young man with blood poisoning in his foot had to have his leg amputated below the knee. Only when he was recovering post-op did he realize that they had amputated the wrong leg. By the time they could get him back in again the shock of the double amputation coupled with the continued blood poisoning meant that he died shortly after the second operation.

The other area of interaction design where left and right play a vital part is in the design of products and systems that depend on which hand the user favors. Most people know about the problems of scissors for left-and right-handed people. If you don't then you are probably right-handed, go and get an ordinary pair of scissors and try using them in your left hand – they just don't cut properly. It is not just scissors, left-handed people have all sorts of disadvantages when using tools and systems set up for right-handers. Even something as simple as the drinks stations for marathon runners can cause problems, they are set up for runners to reach out with their right hand and grab a drink in passing. Tricky if you are doing it at high speed, when tired and using the hand that you hardly use for anything else.

As far back as medieval times the designers (or 'master builders' as they were then) were aware of this difference. Visit any church or castle in Europe next time you are there and you will probably find a spiral staircase or two. Most of them spiral anti-clockwise as they go down (and thus clockwise as they go up). Apparently this was to favor defensive sword use should the building be overrun at any time. If you are up the spiral staircase and are right-handed and facing down the stairs you will have the central axis of the stairs on your left and you will be able to swing your sword at your foe coming up the stairs. He on the other hand will have the central axis on his right and so it will be directly in the way of his sword swings. There is enough contemporaneous information to back up which hand swordsmen used to use in those days, but what about earlier times? Have we always been mainly right handed? Well, there was an interesting discovery recently of cave paintings and hand outlines. As part of the painting ancient man would put his hand on the wall, spatter paint at it and then remove it, resulting in a silhouette of his hand on the wall, surrounded by spatters. Most of these hand prints were of the left hand thus, assuming that the right hand was actively doing the spattering of the paint, we can draw the conclusion that early man was also predominantly right-handed.

As well as the left and right problems with dead bodies that I opened with, there is also a frightening front and back confusion that can arise with half-dead bodies. In the early days of motorbike riding, before they had all the special gear, riders would try and stop the high-speed draughts getting through their coat fastenings by putting their coats on back to front so that the fastenings were on the back. This was fine and warm, the only worry being that if you came off your bike and were lying unconscious by the side of the road some helpful soul might think your head had been twisted right round in the crash and try and twist it back again!

So, left-right considerations are important when you are designing large symmetrical structures with people in them, and they are important when designing things for people to use with one hand. There are probably other situations where a consideration of left-right issues are important but they will only become apparent in hindsight. The best policy is to add a 'left-right issues' check box to your design checklist and consider it, however briefly, for every design project that you work on ... but which side of the text should you put the check box on?

Do what I mean (not what I say)

Hemelsworth: 'Ah, Barker – you've even set up a glass of brandy next to the piano. Sometimes I think that *you* know what I want better than I do.'

Barker: 'Why thank you sir – one does one's best.'

Wouldn't it be wonderful to have a manservant like that; someone who was always one step ahead of you and knew your likes and dislikes down to the finest detail? Well, those days are long gone, but the future holds the promise of artificial systems that know exactly what you want, or at least systems that *think* they know what you want.

These systems can be embodied in non-digital applications and usually they are embodied so seamlessly that you are hardly aware that they are there. In the olden days when you jammed on the brakes in your automobile there was a danger of them locking, so you had to pump the car brakes with your foot to avoid this. Nowadays there are anti-lock brake systems, you just jam your foot down hard and the system does the pumping for you at just the right amount and just the right frequency. A more extreme example is the Eurofighter; the new jet fighter being developed in Europe. The dynamics of the plane are actually unstable, this is designed in because it makes the plane far more responsive and maneuverable. If the pilot was to have direct control, it would probably crash, so the airplane operates by 'fly-by-wire'. Instead of having direct control, the pilots movements are sent to a computer that interprets them and then the computer adjusts the control surfaces itself.

Even something as simple as turning a computer off is tempered with interventions from the system. On the software level you have all the 'are you sure?' messages and the requests to save your files before things quit, and then on the hardware level all sorts of things happen with hard disks shifting their reading heads off the disk and services being disconnected. The on/off button on most laptops itself is tempered with intelligence. One swift poke at the button on my IBM ThinkPad has no effect whatsoever. The system has assumed that I just didn't mean it and that it was accidental. To really switch it off I have to press the button in and hold it in for a few seconds to show that I really mean what I say.

Having intelligent intervention is only possible on a very simple level in everyday products, more complex applications only come about in huge costly systems like the Eurofighter. With the move to the digital world however there is scope for all sorts of interventions from the system. Sometimes, they happen out in the open; the user searches for 'anndroid' and the search engine realizes that there are almost no hits for this and plenty for a similar word 'android' so it comes back with the hits for 'anndroid' and then says 'are you sure you didn't mean to search for android?' In early versions of the search engine when it didn't do this I can remember demonstrating the internet to an osteopath and being disappointed that there were only two documents on the whole of the internet that dealt with the subject. After a few minutes we realized that we were spelling the term incorrectly and had just happened to find two documents that were spelling it the same as we were.

Suggesting alternative spellings is just one way of assisting the user with adjusting the search terms, sometimes it may be better to carry out such operations behind the scenes without the user being aware of what is happening. This is the science of 'cooking' queries. A typical example is searching for names; if I were to search for 'Le Carre' I would get many references to the thriller writer 'John Le Carre'. The problem is that I would probably miss out on finding different versions of his name such as, 'J. Le Carre', 'John Le Carre' or anything with 'Carré' in it. Cooking queries is about not searching for what the user says to search for but searching for what they meant to search for. Instead of searching for the exact syntactical token 'J. Le Carre' you want to search for all possible syntactical tokens that correspond to the underlying semantics of the thing being searched for. Thus I would actually be interested in all variations on his name, including other pen names and his real name of 'David Cornwell'. Of course, the last thing you want is to have a room full of people collating all this, so there are many research projects working on ways of extracting this semantic information from the syntax that is out there – in effect reading and understanding web pages. This dream of 'do what I mean and not what I say' bears little resemblance to the current reality of intelligent intervention which is typified by all sorts pop-up dialogue boxes and little 'characters' that appear in the middle of interactions to offer help and advice on how things should be done. The phrase that is bandied about in the industry is 'intelligent agent'. A more fitting phrase at the current juncture is probably 'back-seat driver'; someone who sits in the back seat of a car and gives inappropriate comments and advice about what the driver should be doing.

Funny Noises

I remember a lesson at school about pollution. As well as the obvious types of pollution the teacher also mentioned 'noise pollution'. In those distant, low-tech days the idea of noise pollution was limited to things like jet aircraft going overhead and traffic noise. In today's world, the explosion in personal technology and the ability of this technology to make all sorts of silly noises means that public places are filled with clouds of gadgets beeping and buzzing.

The first inkling of this problem came with digital watches in the 70s (remember those big, chunky things with one button and a red LED readout?). Well, one of the limited features they offered was a 'beep on the hour' function. In the quiet of the cinema or the school assembly when the hour came, a mini cacophony would sound, spread out around the hour as every watch was set to a slightly different time.

The modern day equivalent of this distracting sound is the mobile phone. A sound as loud and as annoying as an alarm-clock. Especially to commuters woken from a snooze on the train or to audiences of classical music concerts. No wonder they have been banned in certain venues in New York.

Is it possible to list the attributes that make a noise annoying? Well, for starters they are annoying if they occur in a quiet context. Consider going out into the countryside with a squeaky walking shoe, or a context like a crowded train where you want to cut off from the busy world but are constantly drawn to it by mobile phones ringing. Another example is my new oven. A silver thing with knobs in all the right places (well almost) but the oven contains a fan that carries on whirring until it has all cooled down. The result of this is that you cook a pie, take it out and serve it and during the entire meal the fan is buzzing away. Then by the time you get to the after dinner mints the fan has cooled everything down enough and it shuts off. A cooker that is practically designed to make noise only for the duration of the meal is has just cooked!

Noises are more annoying if they are intermittent. Continuous noises are easy to screen out and are less distracting. Noises that occur at random intervals always impinge upon our consciousness more. We can work through buzzing tube lighting, moaning air-conditioning units but if we hear the occasional tings from the heating system or the random hissing of escaping gas from an almost sealed coffee flask it is a continual distraction.

A noise that starts suddenly can give us a shock with the inevitable result of annoyance. How a noise starts is called the 'attack'. If it starts suddenly it can be shocking, if it builds up slowly the effect can be less startling. Sound designers sometimes take this into account and I know of one manufacturer whose telephone makes a gentle 'organic' sounding ringing noise which starts very softly and gradually increases in volume. It certainly doesn't shock the listener, but it does cause great confusion. The psychological effect is not, 'Oh. The telephone's ringing but not loud enough to scare me' but instead it is: 'Oh gosh. What's that strange noise? Oh no, it's getting closer. Is it aliens? What is it? It's almost here… Oh it's only the telephone!'

The owner of a similar telephone with a 'louder and louder' ring-tone told me that his phone sounded similar to an ambulance siren and whenever someone rang up the beeps would start and get louder and louder and the whole family would immediately rush to the window with the kids to watch for the passing ambulance.

Another example to illustrate that difficult to identify sounds are annoying is fridge noise. A colleague has a fridge in the same room that he and his partner eat in, so they chose a special silent model. Silent? Almost. It was quite literally as quiet as a mouse. It produced almost imperceptible swishes and clicks sounding exactly like a mouse nesting behind the fridge. Every time it started up the owners' first reaction was to sit quietly, straining their ears for the noises of a mouse in the kitchen. Even after a year or so it was still catching them out!

A key factor in the frustration of sounds is if you have no ability to control the sound. The worst offenders are bits of technology that make a lot of noise for a long period of time and that cannot be stopped from making the noise. Sounds stupid doesn't it? But that is exactly what today's compact 35mm cameras do. When you take the last shot they realize that there is no film left and the motor cuts in to rewind the film. What they don't realize is that you are in the church taking a picture of your friend's daughter's christening and that the vicar is giving you a particularly hard stare as you try and sit on the camera to muffle the whining sounds.

Does this idea of every gadget making a noise have any advantages? The only one I can think of was the story of the young boy losing his new digital watch while playing football. The football field was so big and grassy that a search by eye would have been impossible. But this watch had an in-built alarm set to go off at 7:30 in the morning. So, early in the morning he and his family and mates spread themselves out across the field and listened and when the alarm sounded they quickly pinpointed and found his watch.

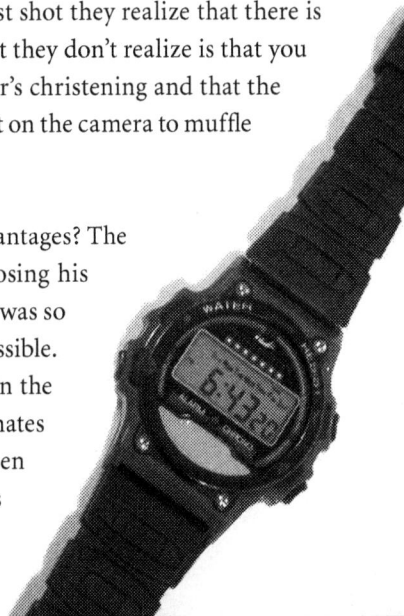

TIME & NARRATIVE

Humans are good at dealing with things arranged in space, we can look around a room, shift stuff around, get from one place to another. We can also design in space; consider the 3D design of physical environments (real or in virtual reality) or the 2D design of flat graphics. However when we are talking about interaction we are also talking about time. We can arrange things in time and design in time but creating a time-based thing (a 'temporal structure') is a lot more intangible than dealing with space. It is the stuff of narrative: storytelling, behavior and movie and theater scripts. The columns in this section look at interaction design issues related to time and narrative.

Firstly, time itself. Time is a key ingredient to interaction, as intangible as it is important. Much of interaction is concerned not only with 'what' is being communicated by the two parties but also 'when' it is being communicated. Think of everyday conversation, a complex set of interactions organized over time with such precision that even a slight error can upset things no end. Just as important as the choreographed to-and-fro of interaction are issues surrounding waiting; both the negative side of it, waiting for the other party to respond in an interaction, and the positive side, politely waiting before doing something that the other party has requested.

Narrative is what happens when other ingredients are mixed with time. It is what happens when something changes and develops with time. This is a complex and fragile topic ranging from simple developments over time through to the way that stories are told and the way that people understand those stories. The columns on this subject deal with general features of narrative like their length, the best way of finishing them off as well as more specific examples of narratives like documentaries.

Snooze Functions

The most delicate time of day for many people is the early morning; trying to match a pair of socks in the dark winter dawn, sorting out a quick breakfast, battling through the sea of commuters. I saw it all nicely summed up by a phrase printed on a cup: 'Please don't talk to me until I've had my first cup of coffee.'

At this sensitive time any contact with obstreperous technology should be minimized. However, at this time most people actually need very obstreperous technology to wake them up and get them out of bed; namely the humble alarm-clock.

I've just replaced my old one. The plastic hands were so old they were starting to curl, not a problem in itself apart from the fact that they occasionally got caught up with one another as they went around. My new one is a bright acid green (which matches the color scheme of our bedroom) but all the hands are black making telling the time in the dark mornings as difficult a task as matching socks.

Anyway, the subject I'm trying to get onto is snooze functions. Recently I wrote a piece about protected functions (Safety Catches, page 42). A protected function has an extra state slipped into the transition from a safe state to a dangerous state. This extra state acts as a buffer state making the dangerous state that extra bit more difficult to reach.

Snooze functions are extra states slipped in between transitions not to protect anything but to make the transition somewhat softer. With the alarm-clock, the simple on-to-off transition is softened into an 'on' to 'snooze state' to 'off' transition. The snooze state keeps reminding the user that they ought to wake up and get their act together. My new alarm-clock, like my old, has no snooze function. It immediately starts beeping and I immediately switch it off, even before I have woken up in the true sense of the word.

If you look around there are other examples of this soft transition between states. The light in cars that is coupled to the car door. It is dark, I open the door to get in and a light goes on. I climb in, shut the door, the light goes off and I struggle to get my seat belt on in the dark. With a so called 'courtesy light' the light remains on for a few seconds giving the occupants time to sort themselves out and settle down. I like the idea that the light is being courteous in giving you a few seconds to sort your seat belt out. My choice of terminology would be to call this a 'pretty obvious function light' and the first example a 'downright rude light'.

Other examples of snooze states fall into the realm of stand-by functions. A sort of 'not dead but sleeping' mode where the system is almost off but is awaiting a wake up signal from the user. Televisions with remote controls have a state where they are not off but are awaiting an on signal from the remote control.

The stand-by state is a case of the transition from off to on being softened with an eye to saving energy. In a similar way the transition from on to off can also be softened with a stand-by state, although here the transition from on to stand-by is automatically initiated by the system itself.

Consider screensavers who's original purpose was to blank out the screen if the system was left for a length of time. Nowadays this function is also performed on a hardware level by energy saving monitors. When I was at General Design we had a PC that had an energy saving monitor coupled to an energy saving graphics card. The two combined to give a stand-by state that was less of a snooze and more of a semi-coma. No amount of mouse or keyboard activity could wake it from its slumber, the only solution was a system reboot. It saved plenty of energy but was pretty hard on the nerves.

The television and the PC monitor are examples of snooze states for saving energy. The alarm-clock and the courtesy light are more interesting – they are examples of snooze states with an eye on the user. The complexity of the interaction is actually increased but the result of the extra complexity is a better system, a system more attuned and better matched to the natural complexity of the user. Hmmmmm. Sounds a bit like neural networks doesn't it?

Waiting

A while back I used to have a workstation on my desk. Click the mouse button on the screen background and you'd get a pop-up menu of options. Due to some quirk of memory, file servers and network, the pop-up menu used to take minutes to appear. I'd go click with my mouse, the hard disk would start to thrash and whirr and eventually a menu would flicker into existence on the screen. Usually I wouldn't be around to see it, I'd have impatiently gone off to wait instead for the photocopier or the coffee machine. In retrospect the term pop-up menu was completely incorrect, a more accurate term would have been a drag-up-painfully-slowly menu.

Menus also feature in La Valade; a restaurant near where I work. You don't have to wait for the menu at all. It's a set meal, so as soon as you sit down they know what you want. However if it's busy you do have to wait a while for the five courses. The funny thing is that the combination of company and atmosphere in a good restaurant means that the waiting is just as good as the eating. In fact it isn't really waiting anymore.

Waiting seems to be a major part of interacting with technology; products and systems are always keeping to the leading edge of what is possible, and often these things are only possible if the system has extra time to do it. Thus the user is always on the waiting end of an egg-timer, a piece of relaxing music or a progress bar. It is the technician's job to keep the waiting down to an absolute minimum but it is the user interface designer's job to ensure that the user feels as good as possible while waiting. The least complex factor in designing for waiting is giving feedback as to the end of the waiting. Many systems do things for users where the activity is invisible, or almost so, and in such cases it is vital to have a clear statement that the task is done.

By coincidence rather than design, our old coffee machine used to signify it had finished by giving a series of gurgling gasps, each one slower than the last. One was never quite sure whether it really was the last gurgle or not. We now have a new machine that is quieter, and instead of gurgling and spluttering it gives a series of sharp beeps when it has finished. The beeps are so severe that for the first week every time the coffee was ready people panicked because they thought that one of the servers had crashed.

Coupled with signaling the end of a process it is also useful to indicate how far in the process the system is (remember the kids in the back seat of the car asking 'are we nearly half way yet?' 'Are we nearly a quarter of the way to being halfway yet?'). Systems should give some indication of the length of time a task is expected to take and an indication of how far they are with the task.

Elevator design is an area where usability designers have to deal with groups of busy people waiting to do a simple task and in a hurry to do it. The progress of elevators can be simply shown with illuminated numbers but this is not exact feedback as someone on the floor above could also be waiting for the elevator to arrive. There are more modern elevators that take this into account and give an estimated time of arrival for the next elevator.

However the question still remains; what do you do while you are waiting for the elevator? And this is the third area; giving the user something to do while they are waiting. You could take the approach of the restaurant and ensure that the time waiting for the elevator to arrive or the copier to warm up was spent in atmospheric surroundings with music and good company, but this would make interactive technology exorbitantly expensive.

One approach is to combine waiting with tasks where the user has to hang around anyway, for example mounting pin-boards or poster boards by the elevator. Indeed one company found an extremely cost effective way to stop the employees complaining about the waiting for the anti-quated elevator system. They simply hung a large mirror up by the elevator entrance on each floor. The number of complaints was significantly lowered.

One nice suggestion I have heard recently concerns waiting in the virtual queue for a telephone service. The queue is virtual but the waiting is real enough. The suggestion was that instead of staying put in the phone queue and listening to the recorded voice saying you are number six in the queue (which is what happens in the Netherlands), what you do is connect the people in the queue to each other, then they can chat (just like in a real life queue) and, who knows, they may even find the answer to the question they were waiting to ask.

Interruptions

A friend of mine was queuing for the check-out in a supermarket recently when there was a fire alarm and everyone had to be evacuated. For my friend this was not a problem, faced with a further twenty minute wait to pay, she was glad to leave the basket there and get out. Which in itself is interesting since it was a way of having the whole shopping experience without actually buying anything.

However, for other shoppers the experience was a bit different since some were in the middle of the actual purchase, they had just handed over cash and were waiting for change, or they had a bag of things but hadn't yet got their receipt. The end-to-end purchase transaction was suddenly being split in the middle. Should they leave the goods they had just bought there in the shop, should the security guards challenge everybody that was leaving with goods, would it be a good chance to grab a few bottles of scotch and saunter out with the other shoppers?

This interruption of transactions is not always a vital scenario to design for. When it comes to sorting out what happens to the people at the till when there is a fire in the shop it is a miniscule number of cases and in each case the transaction is small (on a company perspective). They are certainly not going to re-engineer the transaction system and protocols to completely prevent errors in such a small number of cases.

However, there are situations where transactions can be interrupted more easily and where the transaction is worth a lot more than a basket of groceries. Consider the signing of documents by several parties to purchase a company. Although things can always go wrong in the transaction (person A signs five documents then person B has a heart attack after they have signed the first two of them) the error that has to be prevented is the intentional one. Person A signs a document saying that all their stock in the company is to be transferred to B. Person B grabs it and runs out of the door without signing their document and without making ten million pounds over to person A.

Lawyers who arrange such big deals put a lot of effort into ensuring that what happens is 'watertight' or legally perfect at each point in the proceedings. So that even if the transaction is interrupted for whatever reason, the state of affairs at that moment is complete, legal, and cannot be abused in any way. Legal transactions are interesting because it is human-human interaction that is being carried out according to a very strict set of rules and protocols.

Another bizarre illustration of transaction design in human-human legal interaction is the serving of a notice or a court order to someone in Britain. This is in effect just giving someone a formal document. But what if they don't receive it through the post – how can you prove that they have seen it? If you call at their house and actually hand it to them then they really have received it, but what if they don't want to participate in the transaction and put their hands in their pockets when you try and hand them the document? The legal answer is that it is sufficient to touch the person with the document! (Check out: *www.tbtv.co.uk/stb_social_faqs.asp*).

Managing transactions and coping gracefully with interruptions is also important in the way that pieces of technology communicate with one another. If you accidentally hit the 'mail merge' button on a contact database and the printer starts printing thousands of letters, you want to quickly be able to pull the plug and stop the transaction. What you also want is the transaction to be ended in a coherent way so that the printer will behave sensibly after you put the plug back in. There are certain situations with technology, especially those where you are messing with the more deep-seated bits of programming, where the transactions are very delicate indeed. Installing a new BIOS on an older laptop for example. You have to have a power lead and a full battery and if you interrupt it mid-way you can end up having to bin it because it gets into a state that it is impossible to get it out of.

The early Apple Macs almost got into problems with interrupted transactions with floppy disk ejection. You put a floppy in and the icon appears on the desktop. Instead of doing 'put away' or dragging the icon to the trash can you do 'eject'. The floppy pops out and the icon is left on the desktop but is grayed out. You drop the floppy in the nearest river and next day, on the desktop, you try and clean up by dragging the grayed out icon to the trash can. To dispose of the icon properly the Mac asks you to insert the floppy back into the drive, and - hey presto! - you are stuck in an impossible transition. But those people at Apple obviously thought about this and if you cancel the operation with the apple and period keys, the demand for the lost floppy disappears along with the grayed out icon!

The legal problems described above also arise with the combination of the two; interruptions involving machines and humans. One country had problems with its first ATMs because the legal framework to support them had not been put in place. Legal framework? I hear you ask. Well if I request a hundred dollars from the ATM and then don't actually take it out of the machine but just leave it there, can I legally be said to have taken delivery of the cash? The early machines would offer the money and if the money wasn't taken from the slot the machine would suck it back in again and register the transaction as null and void. Which was fine until someone discovered that you could slip one of the middle bills out without the ATM being aware of it!

Real Time

The World Cup is long over, but there is one match that still sticks in my mind. I was watching it on a large screen TV in a crowded cafe and late in the second half one of England's strikers was heading for the opposition's goal, he got past the first of the defenders and was heading for the penalty area when, in a flash, I knew exactly what was going to happen, I knew that he was going to take a shot and miss, I knew what was going to happen even before I had seen it happen.

What was going on? What quirk of deja-vu or clairvoyance was assisting me? The answer was that it was the crowd further down the room in the cafe, they were also watching the match on another big screen TV but, by some quirk of broadcast technology what they were seeing and hearing was a split second ahead of what we were seeing and hearing at our end of the room. As the striker was approaching the goal we could already hear the disappointed cries from the other viewers and the waning in excitement. At other points in the match we could hear the chants of 'foul' just as two players were running to the ball on our screen and the cheering from the other end of the room just before the goal at our end of the room was actually scored.

Now I know that the speed of light is so fast that the signal difference between one end of that room and the other was negligible so where was the delay coming from? It turns out that it was something to do with modern satellite TV and cables and decoders and stuff, in effect technology is so complex that real-time is no longer real-time. You don't have to find a cafe screening the World Cup to observe this effect, just go into a TV showroom when a large international event is being shown and compare signals on different satellite systems.

What was interesting was the fact that, because of the situation, a minute delay could have such a big effect on the experience of watching the match. Interactions and events are very fragile when it comes to time.

This is apparent on a more regular basis on TV news broadcasts where the news presenter is talking to a foreign correspondent via a satellite link, the round trip for the signal up to the satellite and back again is again miniscule, but it is more than enough to interfere with the normal flow of personal interaction and more than once I have seen news presenters and foreign correspondents stumbling over each others words, then both waiting for the other, then both cutting in again and getting in another tangle. Devotees of the 'nodding donkey' school of philosophy will realize that such a series of mutual silences and interruptions could continue ad infinitum.

If transatlantic delays occasionally interfere with communication then what about the delay in getting a signal to the moon and back; one and a quarter seconds, enough to cause complete breakdown in a normal conversation, no wonder NASA introduced the beeps so that astronauts and ground control could do well organized 'turn taking' in the conversations.

By the time anything gets to the planet Mars, be it a remote controlled robot or a tourist pop-star, the delay for the round trip to communicate with it will be forty minutes. This will make any sort of interaction about as real-time as ancient opposing armies communicating with each other by messengers on horseback.

Interaction, especially technically-mediated, inter-personal interaction, is highly susceptible to even the smallest time delays. Such delays either need to be eliminated or made more manageable by designed systems or communication protocols. And if you move from inter-personal interaction to interaction with technology then you don't have to go to Mars to get delays of forty minutes in an interaction; you can find them here on Earth. Powderham Castle on the South coast of the UK has a guest wing, and possesses such a sprawling, antiquated water system that guests were advised to start the morning by turning the hot-water tap full on and then going for a walk in the ornate rose gardens.

Paths

Instead of lying in bed listening to my bedtime stories Morgan has started to take an interactive role in their telling. For example in the story of Snow White and the Seven Dwarves there comes a point where the Huntsman has the line 'No, I cannot kill you, you are too beautiful, run away and hide in the forest.' At this point Morgan interjects with 'No, tell it where the Huntsman does kill her.' I then have to change to an improvised, grisly, alternative plot that continues, devoid of the main character, and ends with an arrest for murder.

I am telling a truly interactive story, reacting to input from the listener whilst at the same time guiding the narrative to reach a successful closure where I can finish the story with a graceful 'and they all lived happily ever after.' A truly interactive narrative in movie form is the ideal goal of many concept makers working in the media and entertainment industry. We have random access digital media, we have affordable delivery systems and we have a click-literate public willing to engage with new media forms. What we don't yet have is the content; the engaging, interactive movies long promised by the industry.

One reason is that computers are no substitute for a creative imagination honed sharp by improvising nightly stories. One could envisage a true AI based system that 'knows' about the world and how people and things interact with each other and (more importantly) knows what makes a good story. However this is, and has long been, an unattainable goal of more of the industry than just the entertainment sector.

More realistic approaches give limited support to interactive messing with the narrative by ensuring that the choices are thought out and hard-wired into the narrative structure before-hand. This can lead to very controlled interactions with readers all too aware of the points where they must make choices and what those choices are limited to. The Huntsman says 'No, I cannot kill you, you are too beautiful, run away and hide in the forest.' Does Snow White run into the forest? Choose YES or NO.

The latest wave of computer games allows the user to solve a mystery or achieve some other goal by searching a virtual world looking for clues. In effect the environment is scattered with bits of narrative that the user must find and piece together to build up the author's story. This is not an interactive narrative, but more an interactive way of reassembling a narrative. Rather like putting 'War and Peace' through the office shredder and getting someone to try and put it back together again.

What is needed is a way of giving the user the illusion of choice. Either presenting them with choices where the author has got a good idea of what they will choose or presenting them with choices that sneakily lead back to the same storyline. The user thinks they have freedom to choose and the author has freedom to tell their own story.

One of the projects I am involved with at the moment is a website related to a collection of sculptures in a large forest. There are interesting navigational/path following issues in both the sculpture collection and the website. The sculpture trail is a collection of works of art at various locations throughout an area of the forest. Originally the idea was that visitors should explore the forest themselves and 'stumble' upon the sculptures as they did so; a truly user-led navigation with no predefined paths to follow. This led to frustrations: 'Have you seen the Giant's Chair?' 'No, I couldn't find that anywhere!' And some structure was given to the user's navigation through the collection; a pathway was set up linking the key pieces of sculpture together.

Users could now wander through the forest on the path, feeling that they had freedom to step off the path at any point and explore for themselves and yet in reality allowing themselves to be led by the path. The decision to link just the key pieces together meant that there was still scope for visitors to set out on their own paths and discover other exhibits. A further twist to the idea was that unbeknown even to the organizers some of the sculptors added their own mini-sculptures to the forest without telling anyone, so there are some wonderful little lead-cast mushrooms adorning occasional trees waiting to greet those who choose to wander off the beaten track.

Finally, there is the whole issue of real consumer demand for the engaging nature of interactive entertainment. As a follow up to my opening remarks on storytelling, I should add that often, after having jumped through narrative hoops for quarter of an hour at the end of a busy day, I would bid my goodnights and then slump in front of the TV. There I would enjoy a truly non-interactive, non-active time following someone else's pre-baked, mediocre, narrative structure.

```
enter scary forest?    YES / NO
```

Length

Morgan's bedtime stories are about Snow White at the moment. It must always be 'Snow White and the seven somethings': seven kings, seven volcanoes, seven hippos. We have even had some strange, self-referential versions: 'Snow White and the seven Snow Whites' was interesting, and 'Snow White and the seven nothings' was excruciatingly difficult to improvise.

However, one thing that remains fairly constant is the length. If ever I try to pull a fast one and trot out an abridged version, Morgan will be aware of it and complain: 'That wasn't much at all!' Even if I make the story short and incredibly interesting she is still aware that it should be longer. In fact we both have the same idea of the right length, I know this because when I come to 'and they all lived happily ever after' to complete a decent length story she will peacefully go to sleep with no complaints.

This idea of a standard acceptable length or scope holds true for a variety of media: movies, plays, news, music; they all have a typical length.

A single in the charts needs to be five minutes at the most, make it twelve and it gets tedious. I'm always suspicious of thin novels, even if they have won the Booker prize. And consider movies at the cinema. If the movie is a blockbuster spanning seventeen generations of the same wealthy family and it manages to cram hundreds of years into two hours then we can cope with it, more to the point we can live those hundreds of years with the characters. Try and cram it into five hours though and it becomes unacceptable. Despite my suspicion of thin books they are interesting. They are something of a special case as they are not 'consumed' at one sitting so they are not limited by the time a person can sit quietly in one place, because of this there is a wide range of lengths for books. I wonder why the advent of the video recorder and the ability to watch a movie in several sittings didn't have any effect on the length of movies.

It's the same with real life interactions. Each has a fairly standard scope. Someone comes round for dinner; about five hours. Someone pops in for a coffee; about one hour. Game of cards; fifteen minutes. Job interview; 45 minutes. Large variations have a tendency to sound surreal; imagine if someone came round for dinner and had to leave within twenty minutes, or someone 'just popped in' and stayed three days.

Some aspects of the scope of a medium are dictated by the function of the medium. For example there is no recognized size for a telephone book, basically it is as big as is necessary to get all the names and telephone numbers in. Although they do sometimes split phone books up to make them more manageable.

The scope is governed by our limitations, both physical and mental. On the physical side books cannot weigh more than a few pounds (although some of the medieval tomes were massive). On the more important mental side there are limitations governing how quickly we can take information in and process it, and our concentration span. Longer media 'binges' tend to come back to the physical again with the question of how long we can sit still and 'consume' the medium. Although it should be stressed that this is always a balancing act between the discomfort and the interest in the movie. With some movies I have been fidgety and uncomfortable after just ten minutes.

Well, it makes you think doesn't it; just how big should a multimedia CD-Rom be? 650 megabytes you might say – that is certainly the maximum that can be carried – but is that a good guide for the content? It would probably be possible to bind a book with five thousand pages in it, but that doesn't mean it's a good size for books. And the number of kilobytes doesn't really say much about scope and length in time, the 650M could be taken up with one huge TIFF image. And what about a website? How big should one of those be? Functional issues play a part again and maybe it is heavily related to the purpose of the site. The web is a broad medium indeed, so it is akin to asking 'how long should a piece of text be?'

Well as far as the SIGCHI Real World is concerned a piece of text should be 800 words and so that's it for now, and no angry emails saying 'That wasn't much at all!'

Goodbye

First the bad news; this is the last real-world column in its present form. Now the good news; the column continues in the same bi-monthly format on the web at:

www.idhub.com/realworld

And the even better news; on the web it will be supplemented with all the columns to date and other interaction design resources. First addition will be a selection of photos from my interaction design collection; everything from bad navigation in hospitals to ice-cream menus.

And now for this month's column; 'saying goodbye', or more accurately, the bringing to a close of interactions.

Everybody is familiar with the awful feelings left when an interaction has not been finished properly. Having someone hang-up the phone or walk out on you in the middle of an argument. Placing an order on a website and suddenly finding yourself back at the home-page. The worst offenders are movies and TV programs. Spinning their perfectly crafted narrative along until, all of a sudden: Bang! It's the interval. Or even worse; Mulder has finally got through to the secret room where the alien with the funny head is being kept when all of a sudden: 'to be continued...'

Consideration of saying goodbye is important in interaction design because endings and beginnings are vital parts of any interaction. Also, saying goodbye is an important human interaction and designers of public spaces need to pay attention to how the meeting and departing of people are supported in a public environment.

My worst experience of this was the design of a high-speed train that had mirrored windows. I got on, selected my seat and then turned to the window to wave goodbye to my hosts stood outside. The mirrored glass meant that I could see out through a dark-pinkish tinge but from outside it was worse; they could hardly see in at all, and my last impression as the train pulled away was of them waving and staring unseeing into my carriage like a couple of worried looking zombies.

Another key area is the arrivals hall at an airport. People waiting to meet relatives desperately want the earliest peek possible at the person arriving. Normally, you wait by the arrivals doors leading from the customs routes to the public area outside. I have waited at one international airport where there were small windows off to one side between the public waiting area and the baggage reclaim areas. This meant that half the people waiting were not stood outside the arrivals doors but were jostling around these little windows for a first glimpse of their loved ones arriving to get their baggage. Due to a bit of bad design the users were not using the environment as the designers had intended. There was increased traffic between these windows and the arrivals doors, 'he's on his way to the doors quick let's run to meet him.' And a real 'un-designed' bustle around the windows, 'is that him way over there with the green coat on? Quick, lift little Robbie up so he can see.'

In this way the beginning of the interaction with the arriver was scrappily managed, but the ends of interactions also need to be well managed; clarity is vital. Whether it be the decisive bringing to a close of a salesman's call, the formal 'over and out' of a radio conversation or the closing bars of a piece of music; having a clear end informs the other party that part of the dialogue is at an end and that it's time for something else to happen.

The classic example from the world of the telephone is the part human/part computer directory inquiries phone-service where you ring up to get someone's phone number. A human operator talks to you to find out whose number you want and then when the system finds it, a computer generated voice takes over to read the number out digit by digit and repeat it if necessary. What this means for the user is that they begin an interaction with a human and then the interaction is suddenly terminated before they realize it and there is no chance to say 'good-bye', and more importantly no chance to thank the operator for helping them. I imagine that it must be worse for the operators carrying out thousands of such searches and not managing to get a word of thanks from the users.

Knowing about the end of an interaction is also necessary in interactions with systems that have an effect that is not undoable. Making a purchase, placing a bid online, sending an email. The first of these is especially important; you want the user to be able to do lots of things before they place their order (review it, change the amounts etc) and so it must be clear to them which action is the final, ultimate 'really place your order' action.

So it only remains for me to make one final point. The past years of writing this column have taught me many things, but I think the key thing that I have learned in this continual review of real-world interactions is what you might call the 'golden rule' of interaction design and I can state it quite simply and directly: 'The key thing I have learned is…'

To be continued… on a web browser near you.

Blank

One of the worst feelings in the world is receiving an email about an approaching writing deadline and then just sitting there staring at a blank screen on your word processor. There is nothingness and into that nothingness you must bring form, structure, content and above all else entertainment.

Early researchers in user interface design knew this fear of nothingness and at Xerox they quickly realized the power of the copy and edit paradigm over the create-a-new-totally-blank-thing and start-completely-from-scratch paradigm. Instead of having to bring content into the nothingness, you start with something similar and change it until it is what you want.

Nothingness causes a problem in taking that first step in creative or design processes. It also causes problems in navigation and orientation. Having distinct and tangible things in a space is vital to orientation; something to fix your gaze on, something as a starting point or a reference point when building. Arctic explorers have always feared the whiteouts when they were crossing the snow. Whiteouts occurred when white skies, snow covered landscape and clouds of falling snow conspired to create a totally white environment with no perceivable up, down, left or right. A situation likened by some to 'being inside a ping-pong ball'.

In the CAD world this orientation is vital when creating 3D structures. One thing worse than a blank 2D screen representing a blank sheet of paper is a blank 2D screen representing an empty 3D world where you don't even know which way is up and where you are. Some packages solve this by introducing grids and empty but well labeled co-ordinate axes defining the scale and the notion of up-ness that is necessary to begin construction.

However, while the total absence of anything is a thing to be feared, emptiness within structure is a vital, powerful and often underestimated part of the design vocabulary. Just as pauses add drama, doubt, and other emotions to our day to day vocabulary, so too do absences add to the structure of other created things. Within a structure nothingness can form a vital part of the design. Consider the importance of pauses in music; that split second of nothingness after the opening bars of Beethoven's fifth. Or the school drama teacher whose boys are piling into a cupboard representing a prison cell with a political prisoner in it and then immediately banging about as they administer a beating. 'No, no, no!' She says. 'You must all go in, wait for a second and then start the banging, that will be more dramatic.'

In graphic layout the use of so called 'white space' is a key resource, though on the other hand it is wasteful. White space in a newspaper means printing and distributing thousands of copies of bits of nothing at all. On the other hand it gives vital structure and emphasis to key parts of the layout. Architecture too has its own version of white space. Designers such as Rogers and Piano working on the Pompidou Center realized the power in only using half the site and keeping the rest as an emptiness; a clear space which invited others to fill it, ambling crowds, street performers, musicians etc.

The final emptiness I want to mention is the emptiness that seems like emptiness but is not emptiness. Just as the apparent empty space between the galaxies is full of matter so too are other emptinesses actually filled with something. Today's telephone systems sometimes optimize the use of telephone lines and if there is no speech detected for a half second or so they go dead and use the line-bandwidth for something else until you utter another word. This is very efficient, but for a participant in a telephone conversation it can be disturbing. During those vital pauses in conversation you suddenly hear the line go dead... One is almost forced to hum in the gaps to ensure that the conversation is not disrupted by that feeling that you have been cut off.

Well, writing about nothing has ensured that I now have a full screen of something and I am faced with the subsequent writing problem; how to knock all these text fragments into shape to turn it from a collection of bits-and-bobs into a cohesive article...but that is a subject for another time.

Yesterday and Today

Officially the day changes at midnight, but conceptually? Conceptually, it is a different matter. Conceptually, the day changes while you are deeply asleep somewhere when all is quiet and the moon shines between the clouds, somewhere around three or four in the morning. The witching hour! This is the people's view on when the day ends and the new day begins. Needless to say, the view from the technological world is different, and different enough to cause problems. We have a telephone with a voice mail function. Sometimes when we get back late in the evening we check if anyone has called us while we were out. 'You were called yesterday at 21.30' intones the voice mail. Yesterday? Ah wait… Yes it's 12.20 at night now, which means it is not late in the night but early in the morning as far as technology is concerned. As far as I am concerned it is very late at night and I am tired, too tired to work out what the technology is going on about. A sensible person would just say; 'Oh! Earlier this evening at 9.30 someone rang for you.' If I came back at 12.20 at night and the baby-sitter said, 'Oh, someone rang yesterday at 9.30'. I'd say, 'What are you on about, have you gone stark raving mad? You weren't even here yesterday, you've just been here this evening.'

This division between one day and the next is a vital cut off point for all sorts of arrangements of information, it is just a pity that the cut off point is different for technology than it is for people. Another item of technology that gets it badly wrong is the online TV guide that I used to use (before my bookmarks got zapped somehow). It too organized time according to a day that stopped exactly at midnight, even if my own evening carried on for another hour after that. When checking the listings I would select 'today' then click on 'late evening' and then, to see if there were any movies starting later in the evening, I would have to re-navigate to 'tomorrow' and click on 'early morning'.

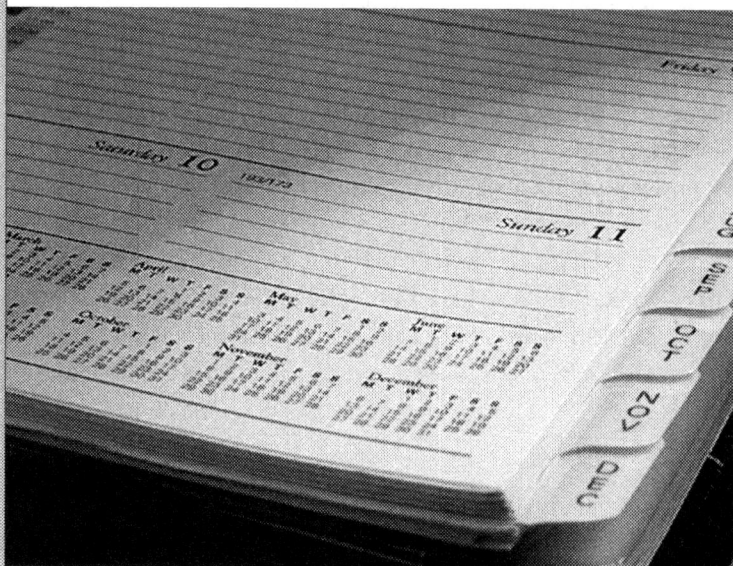

There is one technological system that gets the division right; the synchronization of the speaking clock when the clocks go back or forward an hour. Instead of the change happening at midnight, and causing great confusion to people making late appointments and TV scheduling etc, it happens at two in the morning, which is close enough to the witching hour identified above.

This problem of when technology thinks the day begins and when people feel that it begins is a perceptual problem. But even in the world of technology problems can come about due to the fact that time is different in different regions of what is rapidly becoming a very interconnected world. Even the simple act of ringing friends and family up to wish them a happy New Year at midnight can be complex if some of them are living in different time zones. Midnight isn't a time on the clock, midnight is a long band stretching from the north pole to the south pole that rolls slowly around the globe on the 31st of December.

These time differences and the changing of the clocks by an hour that I alluded to above can combine to yield all sorts of complications. For starters the date that the clocks get their hourly adjustment is different in different countries. For a long time this was done at different dates in different countries in Europe meaning that there were short periods of a few weeks where England was on the same time as its European neighbors and the ex-pats in Holland could set their watches by the BBC news. Even now that things are unified the UK still does it at a different time to our cousins across the pond, meaning that the time difference between London and New York is five hours … most of the time. There is a period of time in spring when the difference stretches to six hours because the clocks don't go forward on the same day, but there is an even shorter sliver of time in the evening of October 30th when the time difference is only four hours because the clocks in London have just been put back but the clocks in New York have yet to do so.

With the current trend of global internet sites serving global audiences and supporting communication between global groups one is forced to ask the question is there such a thing as a global time? Well if you have a look at wide reaching sites such as eBay whose operations do have a degree of time dependence they do adopt their own standard time:

http://cgi3.ebay.com/aw-cgi/eBayISAPI.dll?TimeShow

A final interesting point is the parallel between the internet industry and the rail industry. As well as mirroring each other in the technological boom and bust they precipitated, they are both responsible for influencing the recording of time. The global desire to standardize time brought about by the internet is similar to the national need to standardize time when the railways first started to join up distant towns and cities.

Documentaries

I was walking through the campus of the university with a colleague. We were following the muddy paths carved out of the grass by the trampling students going from one lecture hall to another and talking about documentaries. 'Well', said my colleague, 'this city has quite a reputation for making documentaries. When radio first started we were in there with audio documentaries. In the early TV years we broke the mold with the first video documentaries. Now that everything is interactive I suppose we ought to start thinking about interactive documentaries … whatever they are.'

Is there such a thing as an interactive documentary? Let's start from basics. The word 'document' usually refers to a chunk of information printed on a bit of dead tree. I have documents for all sorts of things, but the word also implies a certain official/factual quality to the information. You have 'travel documents,' and 'legal documents' but you don't have 'shopping list documents' or 'gone to lunch, back soon, documents.'

Add the 'ary' bit on the end and it all gets very different; a 'documentary' is a presentation of this sort of information. More than that, it is a presentation of information woven into some sort of narrative. It tells a story. Sometimes a documentary will actually say that in the title, 'The story of interface design at Apple.' It is about gathering chunks of information, and giving them a structure in time, giving them a story that unfolds.

This brings us back to the old problem of narrative versus interaction. How can we tell a story if it is interactive and the user is in control? This is a question that has engaged the interactive games market for years; how to produce a game that is interactive but that evokes the same emotions as a good story in a movie? It is a real challenge.

As far as documentaries are concerned there are structures where a story can be told yet remain interactive, consider the following four structures:

1) The story of the documentary is built into the system and the user uncovers parts of that story to gradually build up the whole picture. Imagine interactive archeology and searching old records to piece together what happened to the tomb of Rameses the second.

2) The story of the documentary is told from several different viewpoints and the user can switch between them as the story progresses. Imagine interactively switching viewpoints during the story of the Bay of Pigs.

3) The story is built into the context, imagine an interactive simulation where you are a low income worker in India and you get chances to get out of the poverty cycle, gradually it dawns on you that you can never get out no matter what you do. The problems facing the low-income population and the solutions on offer become clearer. (The third-world charity Oxfam produced a board game like this once).

4) The story is encapsulated in a collection of primary-source interviews and archive materials. Rather than editing them together into a TV/audio documentary, they are linked together in a navigational framework that allows the user to explore the whole collection themselves. A high-level narrative element could be included in the organization of the materials. For example: recollections of the invasion of Poland in World War II, recollections of the concentration camps in Poland, recollections of the liberation, recollections of the aftermath and repatriation.

There are other ways, but very little has been tried, and it will be a while before they are produced well enough to succeed.

Making documentaries interactive is a big challenge, but it doesn't stop there. When we get interactive and go on the web we can put a documentary together that is interactive and multi-user. What happens then? Ultimately one could envisage a sort of 'documentary engine'; a multi-user web service that is kick-started with an archive of audio and film and then users react to the archive adding their own recollections, opinions etc and the whole collection goes through some sort of filtering process to extract and present the most visited elements.

Things are moving slightly in this direction, for example the addition of 'talking points' to news in the BBC news website (news.bbc.co.uk). The articles kick-start the process, moderated responses are added from the public, and this becomes part of the document. But isn't the end result just a huge archive of material? Where has the narrative element gone? Perhaps there is some way of letting all the users trace their own paths through the material and then to offer users choices of the most popular paths through it? Users could then follow the popular paths carved out by the hordes of other users.

Similar in many ways to how one follows the paths that have been trodden through the grass in a university campus...

INTERACTION & SPECIFICATIONS

Having considered many aspects surrounding interaction it is now time to take a deeper look at interaction itself. This section examines a few of the vital ingredients to interaction design. Interaction itself deals with a number of abstract design concepts that have to be well executed if the interaction is to run smoothly. There are many simple ingredients to the interaction process, only some of which will be covered here, including putting labels on things and how those labels and other terminology can affect the interaction. More advanced topics include international standards and people flow.

The first columns in this section look at some of the many ingredients of interaction, but having the right ingredients is only part of the story. You also need a recipe to follow. In the world of interaction design this idea of a recipe corresponds to a specification of what you are going to build. A specification pins down the design in a formal way, it says what bits are being used and where they are being used. A specification is a vital tool for the design team and for communication between the design team and the people implementing the design. Here, as well as general consideration of the subjects, we shall also be looking at some more focused examples, in particular formal ideas on shopping transactions and classifying sounds.

part four

Labels

I moved house recently. We did it ourselves and the operation brought many important questions up about the way we live. Are so many possessions really necessary? Do we need to store so many things that we never use? Should we just have possessions that are useful, or maybe just have things that are beautiful? And if we embark on a big throwing out campaign aren't we just adding to the global rubbish heap? Another question that came up, as a user interface designer moving house, was: what is the best way to label the boxes? Initially I started labeling them with a categorization of the contents: 'books', 'cutlery' etc but the contents varied so much that this was difficult. Then I thought of labeling them with the area that the contents had come from: 'mantelpiece', 'kitchen draw'. I added meta-information such as 'fragile' and 'heavy'. Fragile is a good warning to add to a label, but heavy? That would be obvious to the person carrying the box, did I need to add it to the label? Wendelynne suggested labeling the boxes with the destination of the contents, this would avoid huge build ups of boxes at the other end, things could immediately be shunted into the correct places. Problem here was that we then spent ages with each box deciding where the contents should go before we could write the label.

So, a label is about what's in the box, but the question is what's in a label? Well, let's start with the idea that we attach labels to items to help identify them. But wait, I carry my documents around in a cardboard folder in my bag, it's an old folder and bears all sorts of crossed out labels from days gone by. The actual contents are current documents but the label says 'furniture ideas'. This doesn't matter because I never have to look at the label, I only have one cardboard file so I know that it's current documents.

Right, where does that leave us in our definition? Labeling is done when things are in a context of many other things of the same type. Now consider my collection of postcards of famous buildings, they are all the same type and there are plenty of them, but I don't label them, that would be daft because I can see exactly which is which! So, the label is used to convey information about the item that may not be apparent from the perception of the item itself. All right, but when I go to the wood shop they have racks of wood there in different sizes, and I can see it's different sizes and I can measure the sizes but they still put labels on giving the sizes. So sometimes the label is used to convey information that is apparent but that costs time and energy to extract.

These problems with labeling things have parallels in the digital world of the computer. 'Which box did we pack the can opener in?' is the same as 'which file / directory is the beginning of that article that I was writing in?' Labels are vital, but so often they are insufficient. How often have I had to list the contents of a file to see what was really in it because the title no longer jogged my memory? Maybe I should spend longer choosing a good filename each time I create a file. Here we come up against the problem with all labeling systems, be they in the real world or in the digital world; they all require a certain amount of user effort to make the label. Not just physical (typing it in, finding pen and sticky labels) but also mental (what is the best way of describing it? How can I sum it all up in eight characters or less?). Users will always end up doing only the minimum amount of this work necessary to make a label. The Macintosh offers users many labeling techniques including assigning a comment field and a color code to files, but I know few people who actually take the time and trouble to make use of these facilities.

These label creation problems occur in both the real and the digital world, but the problem is exacerbated in the digital world because of an important difference; the real world is richer in detail than the digital, and people use and rely on this richness of detail.

When I'm looking for the can-opener I can't remember what I wrote on the label, but I can remember that it was packed in a long white and blue banana box. I can use the extra detail as a sort of inherent labeling system. On my desk I can find the document about making tables in Web pages easily, not because it has a meaningful label on it, but because it is scruffy looking and has got a big coffee stain on it. I didn't purposefully spend time scruffing it up and pouring coffee on it, it just happened naturally and I can make use of it.

This then is the challenge for the digital world; to create an environment of rich enough detail so that users can bring over the tricks and methods that they use in the rich environment of the real world.

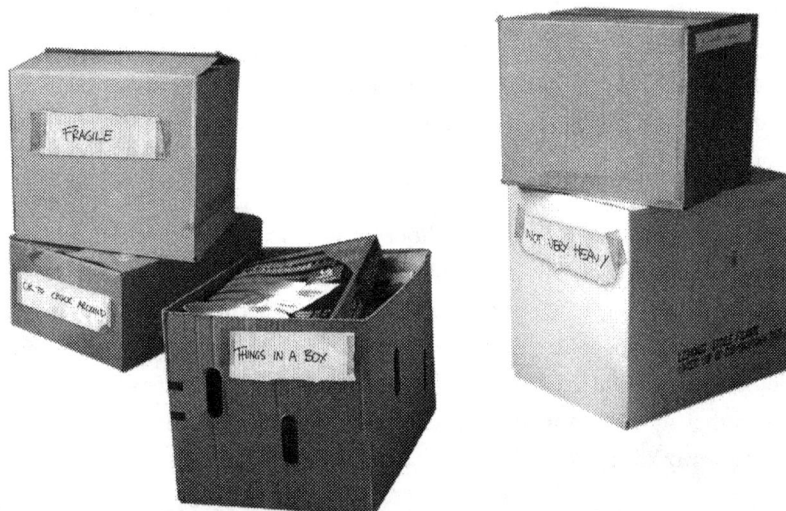

Terminology

I always emphasize that user interface design is not just about the big things like screen layout, user models, information structures and so on, it also includes the little details. Sometimes even the choice of a single word in the interface can greatly influence its use.

The most telling examples come when a wrongly chosen word introduces a new concept. This is fine if it is a concept that plays a part in the user interface, but sometimes a bad choice of terms can introduce a spurious concept that leaves the user confused.

Consider a MiniDisc sound recorder I once used. Each time you recorded you created a block of sound on the disk. The user interface allowed you to skip through and manipulate the blocks with functions like next, previous, play and delete. All very simple and intuitive, but then in the middle of it all was a button labeled 'mark'. Immediately the model became unclear, and questions were raised: What was a mark? What could the user do with it?

Experimenting with it eventually yielded the solution; if you put a mark in you actually split the block into two blocks at that point. Marks were already inherent in the system as the points where one block ended and another started, but they were not explicitly referred to and, as the user model was based on the blocks, a better term would have been 'split block', a recognizable action carried out on the existing concepts.

As a less extreme example the Macintosh desktop interface uses something called a 'clipboard' for that limbo place where things are in-between a cut and a paste, but the functionality is a lot less than the name suggests. In the real world I don't think I've ever used a clipboard when I've cut something out and pasted in into another place. I always hold it in my hand. Still 'clipboard' is a far better choice than 'buffer' which to most non-technical people means the big things at the end of the rails that stop trains crashing. Obviously not a great place to keep important chunks of information!

The people who really invest time into choosing terms are the marketing agencies when they are naming a new product. The choice of the name 'Prozac' for the infamous pharmaceutical product was a huge process. The creatives wanted a name that what easy to say, it had to have an up-beat, positive ring to it, thus the 'pro' part at the beginning. It had to sound modern and technological so they stuck a Z in it (how many software companies stick a Z or an X in their name!) and they wanted an image of progress, of forward motion so they put the 'ac' part in. After weeks of work they created a household name. I sometimes wonder what it could have been, what were the names they tried and rejected? 'Pozgo' or 'Pluzprog', maybe even 'Good-zoom'.

Contrast this attention to detail with the 'Lymeswold' problem (pronounce it so it rhymes with 'times - cold'). A new type of soft cheese made in the south of England with the goal of break-ing into the French cheese market. After several years of poor sales they did some research and discovered that indeed it did have exactly the right English countryside connotations that they wanted. But the problems lay in the fact that the average French person couldn't actually pro-nounce the name, making it very difficult for them to buy it or order it in shops. A clear case of not taking the user into consideration in the design process.

Terms also play a key part in supporting navigational signs. In British hospitals the navigation is based upon the obscure Latin terminology used by the medical profession. The end result is that the doctors who work there all the time understand them (but never need them) while the real clients, the people who are ill, have to spend ages puzzling out which antiquated Latin name means 'back problems' or 'eye department'. There has actually been research carried out on how recognizable the terms are and for some terms less than ten percent of the people understand them.

But my favorite strange terminology was that I encountered while browsing a CD-Rom library index system. All of a sudden I hit upon a paper entitled: 'Graphical presentation using fish'… 'Fish?' I thought, and had this wonderful image of computer graphics labs full of researchers trying to keep piles of trout and salmon cold on the cooling units of a Cray. Then I discovered that the system truncated the titles and happened to have broken this one off in the middle of 'fish-eye projections'!

International Standards

On a computer you can delete directories with reckless abandon. Whole tree structures of folders can be obliterated with just one multiple selection and a control key combination. It's great fun but maybe a bit too powerful at times. Anyway, I have just been doing the real world equivalent; burning real world directories/folders on a real world bonfire, also great fun even if it lacks an undo function.

This may seem rather wasteful but there is a reason. I'll start though with the filing cabinets I bought fifteen years ago from an American town planner. They were 50s styled, rounded corners, Cole Steel, just the right height for a desk top placed on top of them; absolutely perfect. There was only one slight, minor, tiny drawback; they were for paper of the American 'letter' size and here in the UK we use 'A4' which is a slight, minor, tiny bit bigger. I tried A4 paper in the drawers and they fitted ... but only just. So I bought the cabinets. It was a bit like when you find a great pair of second-hand shoes that are just that incy-wincy bit too small, you kid yourself they fit and then spend the next month walking around with curled up toes and blisters. Well I kidded myself about the Cole Steel cabinets and spent the next month with A4 paper that was always slightly crumpled at the edges.

Eventually, I could stand it no more and while living in Holland I switched to European sizes. Faced with a choice of A4 or the larger folio, I chose folio having had more than enough of curled edges. I built up a large collection of information (partly through a policy of never throwing anything out!) and when I returned to the UK I took all the files with me.

On purchasing a filing cabinet to hold all the files I discovered that the UK, like Holland, has two sizes. Unfortunately they are not the same two sizes, we have the choice between A4 or something called foolscap. (I'm still unsure what a cap for a fool has got to do with paper, but there you go!) So I paid my money, got my cabinets, re-filed everything in foolscap sized files and burnt the Dutch folders on the bonfire.

Such international compatibility horror stories are not new. It's not just a paper thing, it covers many other facets of life and it usually crops up at the least expected moment. Everybody knows about light bulbs and electrical plugs. I know I spent my first year in Holland with English plugs on all my appliances and just one plug adapter so that I could only plug one thing in at a time. Having breakfast resembled working in an old style telephone exchange with all the pluggings and un-pluggings necessary to get toast and tea and listen to the radio.

The dangerous things are the more unexpected things. Like the video recorder. It turns out that there is some strange difference in the video signals between the UK and Holland. Meaning that our Dutch video recorder wouldn't work in the UK. The bad usability of this video recorder was the subject of a previous Real World column so I was not sad to see it go. However, at the last minute I decided against putting it on the bonfire and instead gave it to a Dutch colleague.

For the majority of readers living in America, the information here about discrepancies within Europe will not be of much use for those of you traveling abroad. You will sadly be left to discover your own examples! But don't worry, when they do come along you can be sure that they will be as unexpected and illogical as the examples here, and you'll always end up putting something on the bonfire.

The final example is by far the worst and most unexpected compatibility problem. It involves computer software (of course). Not quite as grand as the glitch that crashed NASA's Mars probe but nearly so. A company I once worked for had English and Dutch employees, using English and Dutch versions of a well known spreadsheet program that excelled in most things, but not compatibility. They used the different versions of the program to add figures to one central spreadsheet file (it's called 'increased efficiency through file sharing'). The only glitch was in the visual format of the figures. The English values had a comma as a thousand separator (so; 1,000) while the Dutch values had a dot (so; 1.000). It looked messy but we could live with it. Imagine our horror four or five months down the line when we discovered that there was more to it that just looking messy. The columns being built up in the spreadsheet were actually wrong. The spreadsheet involved automatic column totals and it turned out that these total values only included the values in the English format, all the Dutch values in the columns were ignored, no error messages, nothing! No wonder our accounts looked bleak; we were only counting the values inputted by half the staff, had it been totals of money going out instead of coming in we might even have gone bankrupt.

Now, what's the software equivalent of a bonfire?

Loops

Right from the start of web publishing, standard structures of information were quick to emerge. Probably the most basic of these was the chain; a linear related sequence of chunks of information. Think of stepping through back issues of an on-line magazine, think of clicking 'next' and 'previous' to scan through a collection of photos. The sequence is a very basic organization of information. I do not claim to be an information historian, but I suppose it could have had something to do with the switch from scrolls of parchment to pages. (Although didn't Moses have a sequence of separate stone tablets when he descended from Mount Sinai?).

The shift to separate pages does make information more manageable if it comes with a way of holding those pages together in some way. Without a binding technology a collection becomes difficult to keep in order. Think of flicking through someone's photos in a group of people while the owner shouts 'Try and keep them all in order!'

Organizing information in one big chunk and presenting it as a continuous scroll does have some inherent design advantages though. Scrolls are not bound by the restrictions and meta-issues that have to addressed with separate pages. Issues like size, number, how to chunk the information up. Typesetters even have terminology referring to isolated bits of text on a page, they speak of widows and orphans. Despite this scrolls seem to have had little place in the gap between the Egyptians and today's scroll-bars on computer screens. Now and then they have cropped up; Jack Kerouac used to write his 'stream-of-consciousness' prose with a teletype roll of paper fed into the back of his typewriter meaning that his stream of consciousness would never have to be broken with worries about page breaks or scrabbling for new sheets of paper.

A counter example are those awful fax machines that work with a big roll of thermal paper instead of separate sheets. They are cheap and useful and fine for single page faxes, but when you have received your thirtieth multipage contract you realize that your project archive is starting to resemble a pharaoh's tomb with shelves full of rolls that have to be unrolled for about five minutes to see what they are about! However, you never find yourself in the situation of losing a page from the middle of a vital fax.

The shift away from physical information carriers to the digital world has meant new information structures. Vanevaar Bush was the first to describe the idea in 1945. The simplest of these novel ideas is the loop, unencumbered by having things in piles with a beginning and an end items can just be dealt with in a virtual circle. However the idea of loops is not just something that came about with the computer, there are several bits of good solid hardware where the loop is the basis of organizations.

Remember the 'View Master'? That 3D scene viewer shaped a bit like flat binoculars that was around in the 70s? They took a disk of small slide images, in stereo pairs, arranged in a circle so that once you had clicked your way around the collection you ended up back at the beginning again.

The early days of moving pictures and animation had similar ideas. The zoetrope was a tall cylinder with slits in and a sequence of pictures that would string together as a short animation when it was spun (explaining it in full is not appropriate here). The key thing is, that because the pictures were on the inside of the cylinder they played and repeated as a loop. Thus the most effective animation sequences were those where the first picture was the same as the last and the animation seemed to be a repetitive motion, (it's a similar story with animated GIF images on the web today).

In terms of classic information organization we have the trusty Rolodex. This is basically a collection of index cards mounted in a circle on a central axis and looks like an Elizabethan neck ruff. Spin the handle on the axle and you can whizz effortlessly through the cards from beginning to end and round again.

Finally there is that solid work horse of countless lecture theaters; the Kodak Carousel slide projector. This has a circular slide holder so that when it is full of slides the collection can be looped through endlessly if need be. A brilliant piece of information organization, the only problem with it is working out which way is forward in the collection of slides, and which way up to put the slides and which way around they go and ... but that's another story.

Sequences

The most common structure of things and information is the linear sequence. A collection of items arranged in one dimension with each item adjacent to two others. In England there is that wonderful physical manifestation of such a sequence; the queue. When more than one person is waiting to buy bread, catch a taxi or whatever, they will inevitably form an orderly line and take strict turns. But what are the abstract qualities of a sequence?

If we break down a sequence into its abstract essentials we have three key manifestations:

1) A loop, basically the idea of things arranged in one dimension with items having nearest neighbors. There is no beginning, no end and no direction.

2) A loop with one special item. By flagging one particular item in a loop we can identify a starting point. But there are still two possible directions to go in. A loop with one special item is not useful and has no physical manifestation that I can think of.

3) A loop with two special items. Now things start to get interesting and useful. We actually need two special unique items in order to be able to define a direction. Imagine item A and item B are uniquely identified and are next to each other. Using them we can specify the two different directions A to B and onwards and B to A and onwards.

This idea of direction only coming into being when we have a loop with two uniquely defined items is a bit strange. However, it becomes clearer when we think of a chain (a sequence that is not a loop). By splitting a loop and opening it out into a chain we basically identify two items in it; the item at one end and the item at the other end. Imagining the situation as a chain is a simpler conceptual handle on the same underlying situation as a loop with two unique items. Of course, once direction is possible like this we can include that as part of the specification.

When it comes to applying abstract sequences to the physical world there are all sorts of strange problems. There is the idea of 'next' and 'previous'. There is also the whole science of the labeling of sequences. In the real world we are continually encountering and navigating sequences. From page numbers to floors in a tall building, from carriages in a train to seats in the theater.

Usually we are only aware of the power and ease of navigating sequences when something goes wrong. Take road junctions for example; in the UK, the road junctions on freeways (motorways) are numbered consecutively. This helps with error checking, if you are looking for exit 5 and you see signs for exit 6 then you know that you have gone too far and missed it. By some bizarre glitch of planning, England's M37 freeway does not actually have a junction 6, the numbering goes straight from 5 to 7, completely disrupting the use of the numbers as an error checking device for those that do not know about this glitch. In Holland freeway junctions are known not by numbers but by the area they are in, which is fine in that it gives you contextual information, but it does then lose the sequential information that helps you to spot the errors of missing a junction.

As well as the obvious differences in numbering floors in different countries (I'd be willing to jump from a first floor window in America but not in England) there are strange examples of designed breaks in sequences. The mathematics tower in UMIST in Manchester (UK) used the letters of the alphabet to label the floors (planners and architects tried some weird ideas out in the sixties) someone anticipated problems with letters that looked like numbers such as I and O, so they took all the vowels out of the alphabet (and didn't anticipate any problems with that!). The result is considerable confusion and head scratching for those using the elevators and especially the stairs. On the same theme there is the story (it must be fictitious, surely) of a tall building where floor 13 was left out because no one ever wanted an apartment there due to the unlucky connotations. This didn't cause any big problems until someone tried to do a bungee jump off the top and based the length of the elastic on the number of floors, they overlooked the missing floor and as a result the elastic was twenty feet too long, with tragic consequences.

But back to queuing for bread; in Italy one summer we found ourselves in a very crowded bakers with absolutely no queue whatsoever. After getting to the front and standing there for what seemed like a week we worked out that the order of serving depended on who responded when the baker said; 'next'. The problem was that we weren't sure exactly which of his many words meant 'next'. Eventually the locals at the front realized our predicament and after a hasty discussion and gesturing we were served. In that situation there was a sequence, but it had no physical manifestation. It was an abstract sequence, even more abstract than my opening analysis. It was a sequence distributed through the minds of many customers and servers and augmented with factors such as age, respect, attention, and favor with the baker. A situation which I would never want to try and represent in a software model!

People Flow

The ringing of church bells in England has a long history, almost 4,000 years of it. Much of it is tied up with religion, community and church architecture. However, there are small fragments of it that have something to do with interaction design.

The particular example I am thinking about is the tower in St. Mary's church in Launton, Oxfordshire, England. Here, it was the custom of the bell-ringers to take a barrel of ale into the tower with them to fortify them through the prolonged bouts of ringing in cold and damp conditions. Naturally enough, the vicar took a dim view of these proceedings and when the time came to make some alterations to the church he added a few of his own. As a result the bell-ringing room of the tower was moved up to one floor above ground level, as was the door-way which was reached by an outside ladder. Furthermore the doorway itself was built to be extremely narrow, just wide enough for a bell-ringer to squeeze through and certainly not wide enough for anyone to get a barrel of ale through.

Controlling the passage of people and things through entrances is a key part of the design of built environments, (with parallels in the digital world). The goals of such control are usually restraining traffic to be in one direction or only allowing a particular segment of the traffic through.

A good example of the former is the classic piece of beekeeping technology; the 'Porter bee escape' in effect a one-way valve for honey-bees made of two thin prongs of metal pressing gently against each other, bees can squeeze through in one direction but they can't get back. Isolate a part of a beehive with such a thing and two weeks later all the bees have found their way out and the honey combs can be removed completely bee-free. In the 'people world' such one-way valves take the form of turnstiles to allow people to leave a paying attraction like a museum or gallery without allowing unscrupulous customers to sneak in the back-way without paying.

The other goal, that of allowing a particular segment of the traffic through, goes back all the way to the countryside where there is a great variety of mobile animal life, all of which has to be controlled in some way. The old English 'kissing gates' (no idea where the name comes from) are simple gates that can't be left open accidentally for cattle. Wider gateways make use of cattle grids, these are open grids on the ground whose gaps stop cattle walking over them but allow people and cars to go over. Unfortunately the cattle grid filtering also catches other small-footed creatures such as very young children and hedgehogs. Newer cattle grids have small ramps built in to allow distressed hedgehogs to return to ground level. There are many other examples of this 'filtering' of users. On a domestic scale there are electronic cat-flaps coupled with special cat collars to enable the flap to distinguish between your own little bundle of furs and purrs and whatever those mysterious other things are that squeeze through at night and trash the kitchen. An even smaller example are the 'childproof' tops on pill bottles. The top can only be removed by means of a complex combination of pressing, turning and lining up of arrows. Once again the filtering is not perfect and older people (who are the ones most likely to make use of medicine) often have to enlist the help of twelve year old grandchildren to get the tops off. Road traffic has equivalents with things like speed bumps (which in England have the improbable name of 'sleeping policemen') these punish fast moving cars without harming slow moving cars. I imagine that hedgehogs have little problem with them but one side effect is that they also discriminate against cyclists moving at any speed. The general rule with any of these filtering approaches is to identify the target group you want to stop/slow down and then base the design on features of that target group that are not features of any other group or sub-group.

On the web there are peculiar parallels to the whole science of controlling people flow. If you have a fill-in inquiry form for prospective customers you may want to make sure that it isn't used by students. In e-commerce systems you may want to allow people with an American postal address into the ordering area but divert the others to another area in case they discover interesting discrepancies in the company's global pricing strategy. How can you tell them apart without interrupting their shopping spree with boring fill-in forms? There is a danger that the resulting user experience can end up like suddenly having to fill in a tax return in the middle of Macy's department store.

Such filtering of users is also at the heart of some interactive games, where progress from one level to the next is only made by accepting some extremely difficult challenge. Remember the door from the monastery in the TV series 'Kung-Fu'? The only people who could leave were those that could pick up a red-hot bucket of coals with their bare forearms. No idea how the hedgehogs would cope with that!

Big Choices

In days gone by you would go to the town to visit the store and the storekeeper would chat to you about the things you wanted, he would advise you, help you and suggest new things that you probably didn't know about. This model of agent-assisted purchasing was superseded in 1916 when Clarence Saunders came up with the 'Piggly Wiggly Store'. Despite the strange name it was a breakthrough; it embodied the unheard of idea that the customers should serve themselves! Almost a century later today's stores are massive. They are so big that people have to design the navigation and labeling to help the consumers find what they want. They also have to sort out where to put things in order to help the customer and increase sales of stock.

The problem of finding things in large collections is not just confined to shopping in large stores. It is a general problem that affects many of our activities today. Choosing a video, buying a book, ordering at a restaurant, they are all concerned with making a choice from a large assortment of things. One of the cornerstones of our commercial world in the West is the idea of choice; of giving the consumer (the end user of the designed consumer process) the maximum possible choice. However this provision of maximum choice needs to be combined with conceptual tools for the user to get to grips with the available choice, otherwise the choice becomes overwhelming.

Large choice is not a key factor in enjoyment or user satisfaction. There are systems out there with very limited choice. Consider the conventional TV channels available in the UK, there are only five of them. When I was growing up there were only three of them. Reviewing the options of an evening was a very short process. Now that we live outside the city we have a mobile library van that visits once every two weeks, although it is a large van it still only carries a small number of titles on board. Previously in this column I have alluded to the restaurant 'La Valada', where they only ever have one menu, no choices at all – but that one menu is always perfect and it takes the choosing out of the process and although choosing your food is an enjoyable process so too is just sitting down and talking without having to bother.

Although big choice isn't necessary for a good time it is still a vital part of a service aimed at as large an audience as possible. It is therefore a key problem in designing real world systems and more importantly designing the front-end to internet based systems, where the amount of choice is not limited by physical things such as the size of the shop front or the size of the stock room, rather it is limited only by the limits of the human mind in dealing with the choice. Solve that and you can offer the user a huge range of products. It is a good general design problem to consider and extract rules from, and there are many real-world examples of similar tasks for detailed analysis.

Areas to consider include organizing the items, the labeling of shelves in the store, the classification of items in the on-line store, choosing the scope of what you are offering and what you are not offering. What information about the items to present to the user. Then there are all the user tasks that you need to support; comparing items, keeping track of a group of items before making a final choice from them, budgeting, finding accompanying items, giving the user tools to search and browse the selections.

Finally, we must not forget the idea of recreating the concept of a storekeeper, but this time as an artificial intelligent agent to assist in on-line shopping. The behind the scenes computer systems can gather huge quantities of information about what users are doing and make recommendations to them: 'If you liked this band then you will probably like this band as well.' By selling diverse types of product through one portal they are even able to cross reference between the different products. The last time I was aware of this was while I was choosing a new printer. There at the bottom of the page about the LaserJet 1020 printer it said: 'Customers who shopped for this item also shopped for: The Bone People by Keri Hulme.' So what? Should I buy a copy because I'm thinking of buying the printer? Mind you, I had already read it, maybe the fact that I liked the book means that I should have bought the printer.

Broadcasting

What's the difference between junk email and bell-ringing? No it's not a bad joke, it's a consideration of broadcasting. First, let us clarify 'broadcasting'. One-to-one communication has been around for a long time. Recent technological advances have allowed us to do the same activity but remotely, via post, telephones, fax and internet messaging. What I want to consider here is one-to-many communication. The earliest variations on this were public talking; tribal leaders addressing their followers, generals addressing their troops, and in the field of entertainment there were traveling players and village bards. The interesting things start happening when one-to-many communication goes remote. Then we are on to 'broadcasting'.

Normally we think of broadcasting in terms of media such as television and radio, but the idea of technology assisting in one-to-many communication has been around for a long time before that. One particular example is the bells of the village church. These were used to bring the local community together for all sorts of things, and it wasn't just centuries ago; even as recently as the 1940s bells were still reserved for signaling the alarm in England during the Second World War.

Centuries ago the scope of messages that bells communicated were wider however. They brought the village together and communicated big events such as weddings and deaths as well as signaling festivities and alarms. In a world before time-measuring technology they were also the central timekeeper for village life, dictating the movement of cattle, gathering for religious festivals and a host of other activities in the village calendar. At about the same time other broadcast technologies existed with similar ranges and limitations. Drumming in the African continent. Smoke signals practiced by North American Indians etc.

With bellringing the target groups were very localized (the local villagers) and as such they were well served by broadcasting to a spatially based target group. Everybody within earshot of the bell would fall nicely into the target group. Deaths of local people, festivities, marriages, the people who had lived there all their lives and who would be interested in these things would hear the bells. Those who lived elsewhere and weren't interested wouldn't hear them.

These days the rise in different disciplines, specializations, hobbies, jobs, interests, coupled with the mobility of today's population means that there is only a small interest in things that happen in the locality and people are far more likely to have their informational needs met by other means than listening out for the village bells. TV and radio are today's broadcasting technology and they broadcast to the whole country.

This type of broadcasting really is blanket broadcasting; sending the same information to many millions of people. There is a hint of target groups though; the broadcasting can be adjusted to serve the information only to certain groups of people, but the tools to do this are very impre-cise; basically the time that the program is broadcast and the channel that it is broadcast on will govern who sees it.

With the introduction of the web as a broadcast medium the blanketness and the targeting have both become more extreme. Extreme blanketness is typified by the rise in Spam (junk) email. Email is sent to millions of people in the hope that it applies to one or two hundred of them. If the tide of spam I get were demographically profiled then somewhere I would be listed as a bald, deeply in debt, homeowner who snores heavily during his sleepless nights and needs new ink cartridges daily for his printer.

The wonder of the web is that the technology is there to do the opposite; to target informa-tion very precisely at individual users, so called 'database marketing'. The information then becomes less a case of advertising and more a case of supplying the user with information that is useful to them. The jargon for this used to be 'point-casting'; rather than casting your infor-mation 'broadly' you cast it to a well-defined point, tailor-made for each individual user. The challenge of interactive, web technology is not to give us as much information as possible but to filter us from the deluge of information that is out there.

To return to spam and bell-ringing; imagine if there was such a thing as bell-ringing spam. It would be something like being able to hear the messages of the bells of every church up and down the whole country. 'Mark Bookman has died' - so what, never even knew the chap. 'There's a Spring Festival at Nunney'; no good to me, I don't live there. 'Matchlocke Hall is on fire everyone come and help!'; I've got a bucket ... but it would take me a week to get there.

Information Technology

I.T. stands for Information Technology. The initials have been around for a while now, but what about the subject; when did IT actually begin? A standard response could be: 'Oh in the early 70s I suppose … all the ingredients were there, and the first commercial computers were starting to come into use. UNIX and C were created and so on … Yes, in the early 70s I would say.'

All the ingredients? What are the ingredients of information technology? Well basically I suppose they are 'information' and 'technology', and information has certainly been around for longer than forty years, and technology as well for that matter. A more general definition could be; 'the use of new technologies to manage information for end users'.

Let us try and get a bit more insight by narrowing down to what is probably the key to technology for dealing with information; the link. A link is some piece of technology that allows the user to get from a word or phrase in one piece of information to a different piece of information.

The moment links come up many ITers thinks of Vannevar Bush, and his groundbreaking and easy to read paper 'As we may think'. It's on the web and worth a read, if only to see how ideas before their time are a struggle in the face of a lack of flexible technology to support the ideas. After computers came into the picture Ted Nelson and Douglas Engelbart came along (in the 70s) and brought the idea into the computing world, and gave it a snappy name as well; 'hyper link'. After a few attempts at putting the idea into commercial products and extensive research in erudite academia, the world wide web finally appeared and the rest, as they say, is history.

Probably the most recent development in the saga was British Telecom attempting to enforce a patent of the idea of links in 2000. The original patent was filed after work done on text-based information systems such as Prestel by the General Post Office (GPO) in the UK. The patent was inherited by BT when it was spun off from the GPO in 1981 but they only discovered it in 1997 when trawling through their patent archives. Fortunately for all the case was dismissed in 2002 partly due to video evidence of Douglas Engelbart's hyperlinks from 1968.

Imagine what would have happened if they had got away with it! Every time you clicked on a link you would have to pay royalties. Large information providers would have to offer information structured in an architecture that didn't depend on links - sounds like an interesting design exercise. Thankfully it didn't happen, so the idea is still in the public domain even now.

This idea of linking that BT tried to patent was a very system-oriented way of doing links. The user clicks and the system does the donkey work and takes the user there. But long before this idea there were simpler links where the technology gave the user the link and the user had to follow it up. Think of a table of contents in a book. Probably the first and most useful of these devices was the index, that simple list of words in the back of a book that contains links to the main text.

Even though this has been around for centuries it is still a fundamental and earth shattering idea. Previously, books were linear, they were closed 'black boxes' of information. Not so much information tools as information carriers. This state of affairs was more pronounced in Muslim cultures of the time. Although they had a great tradition of scholars and learning, much of their wisdom was oral (the word *Qur'an* actually means *recitation*) and the printed word was seen more as a communication medium that was part of getting information from one memory to another rather than an archiving system acting as an information resource.

Even in more information-oriented Western cultures around this time the idea of an information resource usually included a human agent, an academic - usually a monk - who had read all the works and had a good idea of what information was in them and where to find it.

With the invention of printing in the 1440s came an explosion in mass production and distribution of information. For the first time information was a product and as part of the process information design started to play a part, initially meta-information in the form of a colophon to a text, but eventually leading on to the introduction of the index in the 1460s.

Think of it! All of a sudden huge blocks of information became accessible without having to read the entire work and even for those who had read the entire work here was a tool that made it possible to dive in and find things that you knew were in there but had forgotten exactly where.

The closed blocks of text – the manuscripts and tomes – were no longer monolithic but were random access, you could jump into the information at any point you wanted in a structured and designed way.

Now that is what information technology is really about.

Specifications

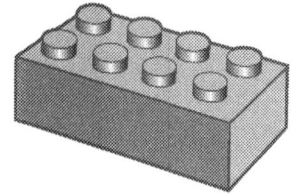

There is a story that the Mathematics building at Manchester University was built back to front. The architect specified it all on paper and then went on holiday (very silly!). When he got back, the foundations were nearing completion and everything was facing the wrong way. It would have cost too much to dig it up and start again, so the specifications underwent minor modifications and the tower was completed back to front. I don't know if the story is true or not but the tower certainly does seem to have a back to front feel about it, and the main entrance is on the second floor.

The specifications that pass between designer and constructor take many forms and the new discipline of user interface designer is still struggling with the ideas. Research groups are continually coming up with languages for describing interactive systems, and yet the commercial world in general is only just starting to take user interface design itself seriously, let alone specification languages for the area.

One of the key factors with specifying interfaces in a real-world, commercial environment is the threshold of effort needed to make a complete specification. Often the approach is to do briefings (rough verbal 'specifications'), and refine them as the construction process continues. This requires regular monitoring, but in the end might cost less effort than meticulous specifications at the beginning. Also, it allows input from the person doing the construction ('Errm, I hate to say this but aren't we building it all back to front?').

One interesting observation is that specifying a user interface could be easy. If you are specifying a digital communication protocol then the requirements (security, error checking) are such that the thing you end up specifying is complex. However, one of the key requirements of a user interface is that it is easy to follow and understand; the concepts must be as few and as simple as possible. It follows then that there should be some simple way of specifying it, and if it isn't easy to specify, then it could be because the interface is badly designed. (If you explain an interface to a group of people and no-one understands it then the chances are that it is a bad interface). So what we need is a language that is close to the designer, that uses high-level, designer-concepts and has nothing to do with how the design is implemented.

The danger is the loss of elemental specification. With any specification language it should be possible to express all possible ideas. Some kiddies construction kits are geared up for building houses, there are large roof chunks, walls and windows. Building any variety of house you want is simple and fast. But if you want to build an airplane then it is almost impossible. In contrast, basic Lego blocks allow you to build whatever you want, but their elemental nature means that what you gain in flexibility you pay for in that it takes a higher level of investment to get the same results. What you need is the best of both worlds; a house building kit where the walls and roofs are prefabricated chunks made from many Lego bricks. You can put houses together very quickly, but if you want to make alterations or create something completely different the large chunks can be deconstructed and you have the flexibility offered by separate Lego bricks.

I opened with the suggestion that a complete specification is often too time-consuming compared to refinement during construction. Another point is that complete specifications are very rarely complete. There are always gaps. Sometimes such gaps are intentional, they form part of a top-down approach to specification, but sometimes the gaps are unintentional and then they are filled in, in one of the following ways:

1) Unknowingly by the implementers; no-one notices the gap and it is just filled in without thinking.

2) According to the implementer's preference; the implementer is aware of the problem but just fills it in with whatever is easiest from a technical point of view.

3) Implementer's guess of the designer's preference; the implementer is aware of the gap, doesn't consult with the designer and makes a guess as to the best design to fill the gap.

4) In discussion with the designer; the implementer (who has already implemented a large part of the solution) draws the gap to the designer's attention and together they sort out a solution, usually resulting in a compromise.

A wonderful example of this last approach is Apple's experiences with Giorgetto Giugiaro the top Italian designer and his studio. After several problems with the joint design process they discovered that the Italian designer's way of producing models was completely different from their own. In America the model shop gets detailed design specifications and follows them to the letter. In Italy the model makers are 'carrozzerie'; craftsmen, and actually take over part of the design process themselves, acting on partial specifications they refine them to produce the final models. The magic doesn't happen in the design studios but in the interaction between the designers and these craftsmen.

Maybe introducing specifications into design always precludes that vital spark of genius that defies specification.

(The Italian designer story is from the compelling book 'Apple Design' by Paul Kunkel, Graphics Inc. NY).

interaction &
specifications

Shopping

'Hello shop!' Me and Morgan, my three-year old are playing shops. One of the reasons for playing shop and doctors and so on is to give children a chance to learn the protocols used in the real world. She is the shopkeeper and I am the customer. As soon as I enter the shop she yells, 'Hello shop!'

After a hasty consultation we try again, this time with me as the shopkeeper. 'Can I help you?' I ask. 'Yes... [pause]'. I realize that she is not going to elucidate, so I chime in helpfully with 'Good, what do you want?' 'Ummm. I want to buy something'. Progress is slow. The next half hour provides concrete proof that protocols are vital to a great many real-world interactions.

The phrase e-commerce is starting to buzz on the internet, the actual interchange of e-money depends on protocols on a system/communication level but the 'shop' and 'mall' systems that use these transactions also use a protocol in communicating with the user. Usually this protocol is very rigid and restricted, leading to frustration for users and frustration for the developers adapting the rigid 'shop' or 'mall' system.

It would be wonderful to have a more flexible system that could support all the commerce protocols that we encounter in the real world. Let's have a look at some of them.

Firstly we have the two key elements to a purchase, the two sides of the bargain; getting the product and paying for the product. Very often, if there is some agent playing a part there is an extra element; selection prior to getting the product. Consider a drinks machine, I put my coins in, press the coffee button and out squirts my drink:

'pay - select - get'

In a restaurant I choose from the menu and only pay right at the end of the meal:

'select - get - pay'

When I buy a train ticket the cashier puts the ticket on one side of a circular tray I put my money in the other and he pulls a lever that spins it 180 degrees:

'select - pay & get (in parallel)'

Is the order: get - select - pay possible? Well I have heard of a cafe where you sit with your coffees and the waiter puts a bowl of Belgian chocolates on the table. When you pay the bill afterwards the number of chocolates you have eaten is totaled in with the bill.

There is also the question of micro- and macro-interactions. In a supermarket what really happens isn't 'get-pay' but something more like:

'get - get - undo-get - get - get - pay'

A chain of micro-gets like this allows for them to be undone. The undo-get is more difficult if the product is a service; you can't undo phone calls for instance (I can recall phone calls that I have dearly wanted to undo).

In principle evaluate is always a part of the protocol:

'get - evaluate - pay'

Or money back if you send the goods back:

'get - pay - evaluate - [undo pay - undo get]'

Moving from cafes and shops to the financial markets we find all sorts of bizarre commercial protocols; options for example. I buy the option to buy a specific quantity of a stock at a given price at a given date; buying the chance to buy!

On a more down-to-earth level I was dwelling on this subject yesterday on the way with Morgan to the cafe La Valade. When I arrived there I was confronted with yet another variation: having sat down, Bo asked me if I wanted a coffee and an apple juice to which I replied 'yes':

'confirm default - get - pay'

And incidentally La Valade serves food in the evening, one set menu (which is always fantastic):

'get - pay'

Commerce in the real world is a very rich set of interaction protocols with a very complex set of constraints. From the systems point of view the protocols should not allow for abuse of the system. (Imagine a drinks machine where you choose your drink, get it and *then* put your money in). From the user point of view there are also constraints of politeness; the user should not be treated like a grade-one criminal. In many European cafes the protocol gives the user constraints priority over the integrity of the financial transaction. You sit at a table, order drinks, get them and you only ask for the bill as you leave. A protocol that also applies to restaurants. Although doing a runner from a crowded terrace is easier than doing one from a restaurant. (Or so I have been told).

Finally, the most wonderfully vague protocol I encountered was in Italy, while visiting a well-stocked butchers we choose all sorts of delicacies which the butcher elaborately wrapped for us, only then did we realize that we wouldn't be able to pay for it all. 'Domani, domani' he said offering us a bit of ham to try: 'tomorrow, tomorrow'. E-commerce eat your heart out!

Sound

The underlying model of a tape for a cassette deck is the same for almost all cassette decks; a physical entity carrying two distinct linear sound blocks. True, a cassette tape does have distinct tracks (songs) recorded on it, but the distinct tracks are not known to the system. It knows only the two sides of the tape and the beginning and end. This rather monolithic model is changing as technology improves the underlying model. CDs and CD players know about tracks and can do interesting things with them. But the era of cassette decks was interesting because, although the underlying model of the tape was the same, different decks provided different controls to the user. These controls embodied actions and the key thing was that the user's goals were achievable with these actions. Consider 'rewind'; standard decks offered a rewind action. Contrast this with in-car cassette decks where the cassette was inserted sideways with quite a large part of it sticking out. These decks only had the actions 'play' and 'fast-forward'. To achieve the user goal of rewinding a cassette, a composite action was required, namely; 'turn cassette over' and 'fast-forward'.

Another less common example is the child's cassette deck that my daughter has (and that I frequently borrow for lectures!). To achieve the goal of playing at different volumes it doesn't have a play action and a volume control it has three distinct actions/buttons; 'play quietly', 'play normally' and 'play loudly'. Different actions but the same goals are achievable. Sometimes the technical limitations mean that the user is obliged to carry out tricky, composite actions in order to achieve simple goals. Finding the beginning of the next track using the

actions 'play' and fast-forward' is possible but very unwieldy. With a video recorder it is even worse, as the video recorder needs to do a lot of whining and clicking between each change of operation.

Nowadays the underlying model has been improved with the introduction of sound technologies such as DAT, CD and MiniDisc. These support the idea of separate sound chunks, yielding improved actions (finding the beginning of the next track is a doddle) and a host of new actions like 'shuffle' (play all tracks in a random order) and 'play intros' (play just the first five seconds of each track). The 'play intros' action migrates well to other sound-chunk technologies such as skipping through voice mail messages, while applying a 'shuffle' action on your voice mail would definitely not be a useful operation. The most recent, interesting development in this area is the 'digital wallet'. A box that is, funnily enough, about the size of a cassette tape. It can store several gigabytes of sound tracks (or photos or anything else digital for that matter). Here you can store not just your favorite tracks for the day but your entire music collection; that whole shelf of CDs in one unit. This is interesting because it breaks down the entity of 'an album'. You can apply a shuffle function to your entire music collection and get all sorts of strange combinations of tracks from different artists and different genres. It's something like having your own radio station playing only music from your own collection.

The final gem to come out of this overview of music technology is the idea of 'technology overload'. This is the practice of extracting advanced features from media whose underlying model does not support those features. Dictaphone users spent ages struggling with the problem I alluded to earlier – of finding the beginning of tracks – until one manufacturer introduced a system where the user could press a button to insert a beep as a marker onto the tape, and had actions to skip between these markers during playback. This is a simulation of the separate tracks that are an integral part of the underlying model of formats like CD and DAT. Another example were the cassette players that searched for the start of tracks by playing through quickly to themselves and 'listening' for the gaps between tracks. Again imitating the separate track formats of CD and DAT, but this time not extending the format with user-inserted markers but applying low-level AI to the existing format. However, like all low-level AI it was not 100% perfect, and the cassette system I had as a student played havoc with my Joni Mitchell tapes, turning some of her solo voice tracks into multiple tracks with just one verse in each!

Forms

For me, and many other people, the filling in of forms is one of the most gut-wrenchingly difficult tasks that there is. Tax forms make me cringe, and forms connected with nasty things like car insurance claims are frightening. Often the problems associated with form filling are to do with how well they cater for your particular situation. If they match well you can just get on with it, if they match badly you spend all your time filling things in and crossing them out; 'do they mean this or do they mean that?'

A small part of good form design, be they paper or on-line forms, involves guiding the user through the filling in of the form; only asking them questions that are relevant and hiding information that is not relevant. However a deeper part of form design is the building blocks at the foundation of the very system that the form is a part of. This is the aspect I want to concentrate on here; forms as an indication of how closely systems match reality. A form that needs to be filled in is not just a means of gathering data, it is an embodiment of assumptions made by the system (and the system's designer) about who the user is and what they are doing. Forms are the 'skin' of underlying systems, and systems are often set in old ways of doing things and old ways of classifying people that don't match the real world.

Here is a common example: although a huge proportion of long-term, stable relationships do not involve marriage, there is still very little recognition of this in forms and processes. For men they must either tick 'Single' or 'Married', there is no box for 'Actually living with someone for the last twenty years and fathering their kids and probably going to be with them a good sight longer'. For women its even worse, if they have already been married and then got divorced before settling down without marrying, then for the rest of eternity there is only one thing they can choose when faced with the choice: 'Married, Single or Divorced'.

Even the simple, multiple-choice questions I've been working on recently for an on-line, user survey have had to be adjusted to take all the user's eventualities into consideration. As well as the five optional choices, two extra options have been added. An 'other' option in case they have an answer different to the five choices presented and an 'absolutely no idea' option just in case they really don't have any idea what the question is getting at.

The bottom line is that it's difficult to design a form or a system to cater for every user situation; you just have to make sure that you cater for the greatest percentage and have a way for users outside this group to also express something. The more loose and un-rigorous a particular business process is, the more there needs to be scope for the user to express their situation in a way not captured by the structured interface of the form. In these situations paper forms come with plenty of white space for additional information and interactive systems need plenty of event logging with accompanying free-text comment fields to catch non-rigorous, user eventualities.

By far the best example of a 'non-rigorous, user eventuality' was Jane's dad's car crash. His car had a problem so he left it with his cousin who was a car mechanic (and scrap dealer!). At the weekend he and his family were watching the 'Late Late Breakfast Show' on TV, and the TV company were filming a live stunt at a scrapyard. The stunt involved dropping one scrap car from a crane onto another scrap car. Suddenly it dawned on Jane's dad that the scrapyard was his cousin's, and that for the shoot the TV crew had chosen two of the best scrap cars that were there, and one of them wasn't scrap at all it was Jane's dad's Ford. While he tried desperately to phone his cousin at home and the TV company, his family shouted a running commentary from the lounge as the scrap car was hoisted up and dropped from the crane onto his trusty Ford.

Looking back he can now laugh about it, he says, but the funniest thing was filling out the car insurance forms. A true indication of how forms and business processes can never capture every single user eventuality, and probably the only time someone has had a laugh filling out an insurance form!

How fast was your car travelling?	STATIONARY
How fast was the other car travelling?	SEVERAL HUNDRED MILES AN HOUR
How many occupants were in the car at the time?	NONE
How many occupants were in the other car?	NONE
From what direction did the impact occur?	FROM ABOVE
How many witnesses were there to the accident?	ABOUT 5 MILLION
What are their names and addresses...?	

Flexible Systems

'No one has ever asked me that before'. What was I asking? Well it was quite simple really. On Monday I had bought some flower pots and compost at the garden center, and arranged for them to be delivered on Friday. Fine, that's part of the service they offer. But what I had done then was to go back a few days later and buy some more stuff, chairs this time, and ask to include them in the delivery that was still waiting to happen on Friday.

This was not part of the model of their service. If I had just arranged a separate delivery of that second load of stuff that would have been fine, but I would have paid for the extra delivery and they might have dispatched it as a separate delivery. Merging it with the first delivery would save me money, make their delivery process more efficient, and it could also benefit their sales since being able to add more to my impending delivery is likely to entice me to buy more big, heavy stuff.

As it was, although no one had asked such a thing before, it was a simple matter to dig out the old order and add the new things to it with a pen and then put the things in the holding shed with the other stuff. But the key thing was that this was an informal solution to the problem; Mike had to get Frank on the phone and they sorted it out between them, it was possible to grab a pen and add a few more things to the delivery schedule etc. With real-world processes like this you can do that, when the organization gets bigger and when things get more rigorously set in processes it becomes more difficult, and nowhere is this more apparent than when a process is automated by computers.

I recently had problems with the acceptance of my credit card for payment because the system would only approve the transaction if the credit card details matched those in the database and if the address details also matched. This meant that the person I was speaking to filled the details in and then pressed on a button to make the system try and match them. It then came back with a no or a yes. As my address is a bit vague and as the operator couldn't see the

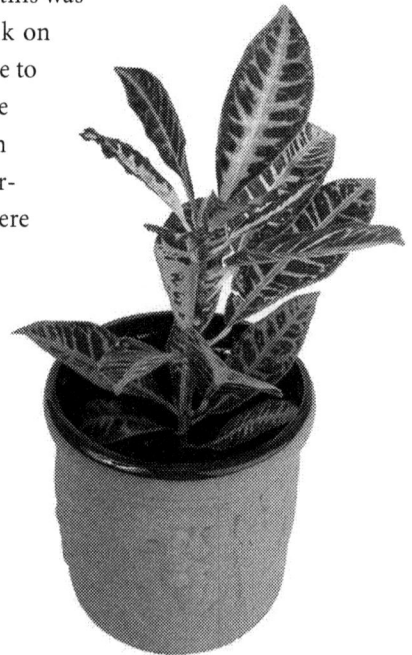

information in the database we had to try submitting several variations before we managed to hit upon the version that was actually stored in the database and allow the transaction to happen. Now if Frank or Mike had been there with their trusty pen they would have flicked through the filing cards, found the matching credit card number and when I said 'Manor Cottage, Norton Malreward' instead of 'Manor Cottage, Church Road, Norton Malreward' they would have said 'That's near enough for us, thank you'.

The lesson is that when you are automating a process in an interactive system you obviously have to think of the main process that you are supporting, but you must also think of the other possibilities, and work out which of them you are going to support and which of them you are not going to support.

One way to avoid this or at least to postpone solving the problem is to include free text comment fields in key parts of the interface. This allows users to annotate things and can also be used retrospectively to catch parts of the process that were missed in the initial requirements drafting. Consider the comment field associated with Macintosh files, although not widely used, when they are used they are used to support all sorts of different processes. It is the digital equivalent of a Mike and a Frank and a quick fix with a pen.

Moving processes from the real-world to the digital world means that we lose some of the advantages in flexibility of the real-world; the Franks, the Mikes and the quick changes with a pen. However, we gain much in speed and efficiency by getting rid of some of the problems associated with paper processes in the real world, one such is the ability to lose key bits of paper down the back of cupboards and such like. One large medical insurance company that had experienced problems of this nature installed a company-wide, digital system with a gateway between the real-world and the digital world consisting of a post room where every paper communication that came in was immediately scanned into the system with a date stamp. This gateway room was a clean room with just post trays and a bank of scanners on simple tables. No objects or cupboards down the backs of which the Mikes and Franks of this world could lose forms.

So when you are transferring processes from the real-world to a digital system you must ensure that you migrate the useful things from the real-world to the system while getting rid of the non-useful things.

And incidentally, this wasn't the original subject I had planned for this month's column, all I know about the original idea was that it was really good and that it was written down on a bit of paper which got lost down the back of something in my study. It will turn up one day…

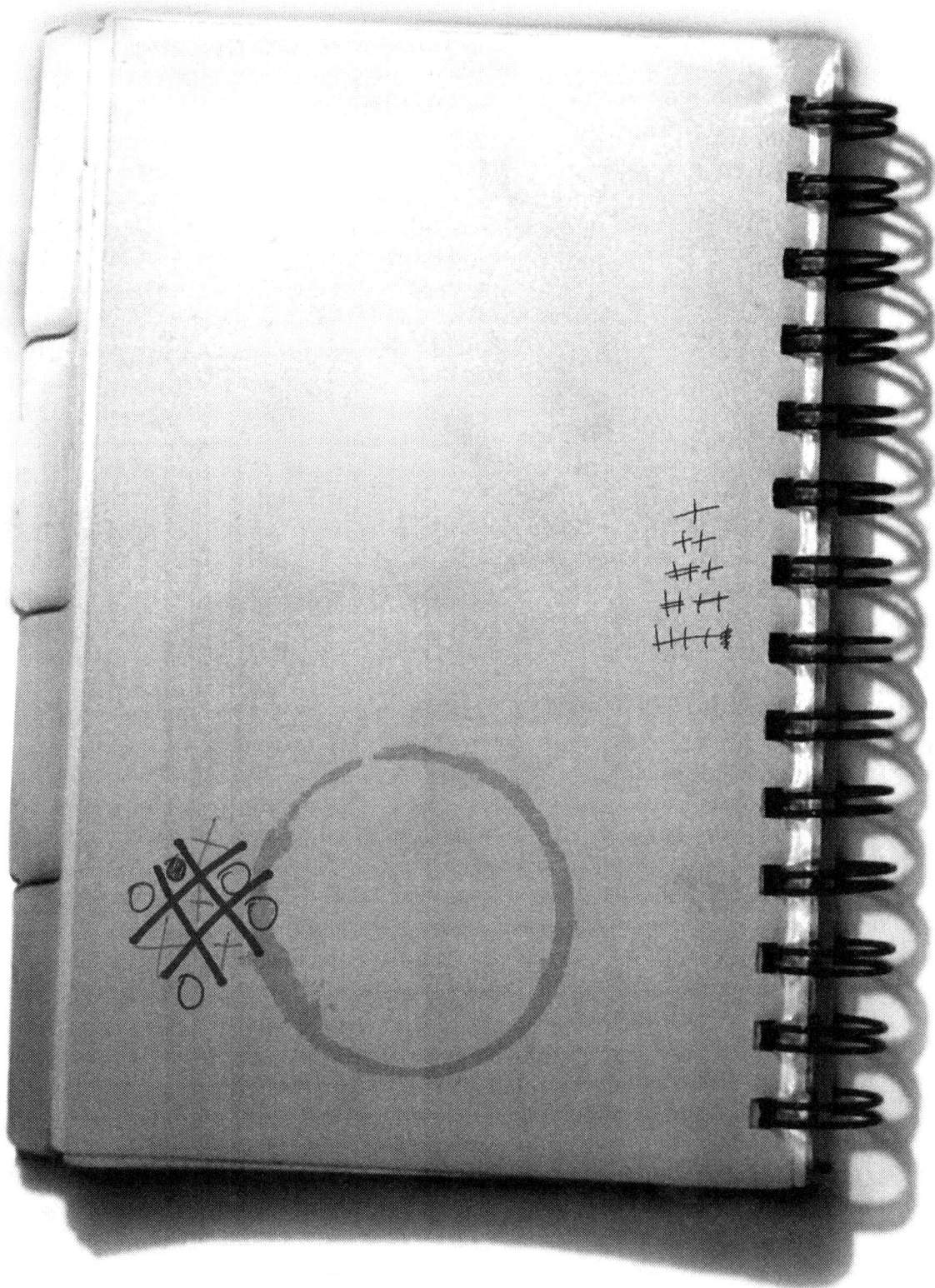

CONCLUSION

Ten years worth of interaction watching. What can we learn from it? Is there a future for increasingly sophistication interactive systems and can our interactions in the real world feed the effective development of such systems? Looking back through this collection of fifty columns what conclusions can we draw?

Life is interaction

The psychology of interaction is fundamental to every aspect of our lives. Basically our lives are interaction, it could even be fundamental to some definition of what life is. The more interaction there is, the richer our lives are. The less interaction, the poorer we are. Everybody interacts, from the chatty socialite to the monk with a vow of silence tending the garden. From the politician 'pressing the flesh' to the reclusive hacker sat behind his terminal.

Design affects our lives

Good design improves our lives and bad design degrades our lives. Good design is design that is good, not design that looks good. There is a big difference.

We live wrapped in a cocoon of designed products and services. Two of the main factors governing the design of these services are the cost and the user. As companies struggle to keep costs down they also struggle to maintain or improve the user experience; it is a difficult balance to maintain. Any costs they can save can be invested in more products and services for the user's. Perhaps we are sacrificing quality for choice; instead of having access to a handful of quality interfaces we have access to hundreds of bad or mediocre interfaces. Either way, the design of the users world is important to their well-being. With the West's preponderance of designed elements in the environment and its lack of other established sources of well-being (peer respect, religion, position in community) the contribution to well-being played by the designed environment becomes vital.

Good design is not everything
Life is complex. Good design improves life but it is only one of the many things that improves life. Good design will make a difference in everybody's lives, but for some people it will make a bigger difference because there may be other things lacking in their life. Good design is no substitute for a broad range of life-improving factors.

If you want to improve someone's life through design then design interactive things well, but if you just want to improve someone's life, there are many other ways of doing it some of which are far more powerful than designing technology.

Interaction design has come of age
Interaction design is now part of the establishment, it has come out of the labs and into the homes. Big corporations invest money in it, users are aware of it, consumer groups test for it when comparing products, people in the street know what you mean when you say that something is 'user-friendly'. When users have problems with products they now sometimes blame the designer instead of themselves.

It is a good time to be an interaction designer. There is lots to be done and there is an underlying support for doing it, both with the public and in industry.

Be open
Examples, such as those in the columns, can only be garnered and used if a designer is willing to keep an open eye and an open mind at all times.

Interaction design as a subject is developing and changing. Interaction design as a discipline with a place in the commercial world is also developing. Progress will only be made if there is insight into what people are and how they work, and that insight can only be found in people who are willing to learn, willing to appreciate the nature of this new discipline and willing to be open to observing and learning about the rich world of interaction in all its guises.

A P P E N D I X   A

R E A L - W O R L D   I N T E R A C T I O N   F O R M S

WATCH AND LEARN

I have already mentioned that this book has been designed as a resource; the designer of inter-action can read through it, peruse it or actively reach in for examples through several indexes. It is a collection of observations that can be used in a variety of ways.

However, that is not the main thing that this book is about. This book is less about me looking at the real world and more about you – the reader – looking at the real world. It contains sup-port on the process of looking to the real world and the columns are also intended to inspire you to go and collect your own examples. Open your interaction designer's eyes and get out there!

In this appendix are a few forms to help the process along. If you are reading this book while waiting for a train or plane then start wandering around the terminal and fill them in as you wander. If you are in the library then run off a few copies of them and keep them in your ruck-sack. If you are teaching then give them to your students as homework. Have a small pile in the kitchen or in the office as therapy, so that you can at least salvage some useful observations on those occasions that interactive technology exhibits bad design … as, sadly, it so often does.

Alternatively, fill in our on-line form at *www.idhub.com* to get your observations included in our next book.

NOTES: Observed Real-World Interactions

Date:............................ Location:..

Description of System

(eg. ATM)

Description of Interaction   GOOD ☐   BAD ☐

Abstract Issues Involved

(eg. Navigation)

NOTES: Observed Real-World Interactions

Date:...........................Location:...

Description of System

(eg. ATM)

Description of Interaction   GOOD ☐   BAD ☐

Abstract Issues Involved

(eg. Navigation)

NOTES: Observed Real-World Interactions

Date:............................Location:..

Description of System

(eg. ATM)

Description of Interaction GOOD ☐ BAD ☐

Abstract Issues Involved

(eg. Navigation)

NOTES: Observed Real-World Interactions

Date:...........................Location:...

Description of System

(eg. ATM)

Description of Interaction   GOOD ☐   BAD ☐

Abstract Issues Involved

(eg. Navigation)

APPENDIX B

SUMMARY OF COLUMNS

As well as the conventional index at the end of this book, I am including this list of the columns detailing what each column is about and what examples are covered. This should be a useful way of finding the content you want or of locating an example when you have forgotten which column it occurred in. This list is ordered as the columns occur in the book. Readers interested in the chronological order in which the columns appeared should refer to appendix C.

PART ONE: HARDWARE

Buttons [page 32]

Many interactive devices have buttons. What effect does the interaction design of the buttons have on the users behavior with them? How do things change when we consider the design of on-screen buttons?

(IP numbers, hi-fi systems, digital alarm-clock, digital watch)

Slider controls [page 34]

As well as discrete on/off buttons there are controls that gradually change something, think of the volume knob on a radio. What are the key issues with these controls and what does the user want from them?

(Gas cooker, volume knob, car doors, TV presets)

Volume control [page 36]

Appliances that make sound also let the user control the volume of that sound. However, there are many ways of doing this and a surprising number of problems can arise with this simple action.

(TV mute, sound card, laptop volume)

Off and on [page 38]

It is important that the user knows if something is on or off, or just not plugged in yet. What feedback is used to communicate this? What about the different levels of 'being on' for computers? And there are always things that may have just been switched off but that are still hot and dangerous.

(Electric kettle, toaster, Sun workstations, video recorders)

Powerful functions [page 40]

What are the pros and cons of having one key press to do something intelligent and powerful such as the 'last number redial' button on telephones? In computers there is the 'copy' function, so easy and powerful but with untold problems.

(Telephone buttons, large file systems, 'open book test')

Safety catches [page 42]

In special situations certain actions have to be possible but difficult to get to. There must be no chance of them being done by accident, or in the case of 'dual key control' no chance of one deranged person doing it on their own.

(Starship self-destruct, fire extinguisher, 2001: a Space Odyssey, saving files)

Coffee [page 44]

Complex coffee machines in large organizations can be wonderful illustrations of common interaction problems.

(Coffee machine)

Telephones [page 46]

More and more complex and novel interactive phone services are being developed, what sort of impact can they have on the ordinary end user?

(Last caller number, voice mail, call redirect)

Elevators [page 48]

The average elevator is a complex multi-user interactive system that provides all sorts of insights into interaction design. From button pressing to information presentation.

(Elevators 'ding-dong' signal, information presentation in elevators, floor numbering schemes)

Video [page 50]

Some video recorders embody intelligent assumptions about the user's tasks. However these are not always apparent to the user! What tasks did the designer have in mind?

(Video recorders, automatic functions of video recorders, TVs)

Video conferencing [page 52]

A user interface designer takes part in an international video link up using state of the art technology, has a really good laugh about it and draws parallels with what it says about the experience of disabled people.

(Thunderbirds, video conferencing, telephone on-hold music, disabled users)

Cafés [page 54]

Cafe design is a discipline where the user experience is the key to success or failure. It is made up of many factors from location to interior décor, and not forgetting the quality and price of the coffee.

(Cafes, coziness, nature, trendy chairs, wobbly tables)

Food [page 56]

Interaction plays a part in running a restaurant, there are problems that cooks and waiters have with usability in these high activity environments.

(Noisy environments, pizza orders, garlic, 'white sauce syndrome')

Infra-red [page 58]

Remote controls have gone from cables to ultrasound to infra-red. There are interesting issues surrounding the use of infra-red as a remote control technology.

(Ultrasonic remote control, laptop IR communication, IR car locks, stealing IR codes, zapping moths with ultrasonic sound)

PART TWO: PEOPLE

Contact [page 62]

New technology makes it easier for people to contact one another, is it reaching the stage where everyone is contactable all the time? There are parallels with the impact of teenage car ownership in the fifties.

('The Graduate', mobile phones, ICQ, WAP, love-getty, hot badges, romantic faxes)

A user group of two [page 64]

There are a growing number of systems that are set up for two different users, each user having their own personal preset. What advantages do these systems offer and can the ideas be generalized to other technologies?

(Toasters, alarm clocks, car seat presets, duvets)

Junk [page 66]

I like many others have plenty of junk in boxes in the real world. How do we deal with it and are there any parallels with junk in the digital world? All those folders and zip disks with copies of copies of copies of files on them.

(Storage units, Eiffel tower rivets, digital desktops, tidying processes)

Marks and scratches [page 68]

Certain pieces of technology last a long time and get scratched and beat-up and almost become fashion accessories. What is the appeal of beat-up technology?

(SLR cameras, archeologists trowels, Psion organizer, living space, doorbell tunes, ring tones, WOOL window system)

Take-out service [page 70]

In public services there are many parts of the service that can be stolen. What are the different approaches to stop people walking off things?

(Shoe shops, sleeper trains, cups with holes in, spoons with holes in, customized clothes hangers, headphones)

Stereo vision [page 72]

Our eyes and ears are adapted to gauge position in a three-dimensional world, but sometimes they can get confused.

(Flatland, shouting from balconies, the Cyclops, random dot stereograms)

Being overheard [page 74]

With extra communication gadgets around, the ability for others to overhear us is increasing. What sort of problems can this bring?

(Conference phones, baby monitors, political gaffes with open mics, old ladies with video camera, hotel chandelier, Furbies)

Children [page 76]

What lessons can be learned form children interacting with technology and from children's toys?

(Language convergence, TV controls, rugby, clockwise/anti-clockwise, visual filtering, importance of sound)

Pointing [page 78]

What is happening when we point at something with our finger, and how does this compare to pointing at something on-screen with the mouse?

(Balloon help, buttons, mouse usage)

Left or right? [page 80]

There are many cases where confusion can arise between left and right. What characterizes these situations and are there any other design issues to do with left and right?

(Dead body, airplane crash, tour guides, leg amputation, scissors, marathon drinks, spiral staircases, cave paintings, motorbike jackets)

Do what I mean (not what I say) [page 82]

As systems become more intelligent, they have more ability to interpret our commands instead of responding to them directly. They can make decisions about what we are really trying to do and can try and do it for us.

(Anti-lock braking systems, Eurofighter airplane, computer shutdown, search engines, intelligent agents)

Funny noises [page 84]

Our world of technology is now awash with sound. Beeps from watches, unwanted mobile ring tones and noises from kitchen appliances.

(Noise pollution, beeping watches, ring tones, kitchen oven, 'organic' ring tones, fridge, automatic film rewind)

PART THREE: TIME AND NARRATIVE

Snooze functions [page 88]

Alarm-clocks that can give you an extra ten minutes in bed are well-known, can the snooze idea be generalized to other technologies that don't do things at once but give you a few seconds grace?

(Alarm-clocks, courtesy lights in cars, computer stand-by states)

Waiting [page 90]

Waiting is a common part of daily life, waiting for computers to react, waiting for food to arrive in a restaurant, waiting for the elevator to arrive. How can the interaction designer deal with waiting?

(Pop-up menus, restaurants, coffee machine, elevators, interaction in queues)

Interruptions [page 92]

It is always possible for an interaction to be interrupted in some way. How should interactions be designed to handle interruptions and what happens when an interaction is prematurely brought to an end?

(Supermarket fire, signing legal documents, serving a court order, mail-merge, ejecting Mac floppies, ATMs)

Real time [page 94]

With every sort of technology there are delays in real-time transmission. How do they effect the interaction?

(Real time TV, satellite links, space communications, plumbing delays)

Paths [page 96]

The narrative path followed by a story can take different directions depending on how it is being presented. The user can be provided with simple options or with real improvisation. Are there parallels with following paths in a landscape?

(Children's stories, forest sculpture trail)

Length [page 98]

There are accepted lengths for various interactions/presentations, everything from a dinner date to a feature movie. How does this tie in with the function of the interaction and what does it tell us about the length of new media products?

(Children's stories, movies, telephone directories, CD-Roms)

Goodbye [page 100]

This was the final column from the print version, it was all about finishing interactions or continuing them later.

(The X-Files, train departures, airports, directory inquiries service)

Blank [page 102]

The use of blank space in layout and silence in sound design is vital. Pauses in plays and movies that can make or break a scene. Can we generalize about the use of nothing when we are designing interaction?

(Blank page, Xerox PARC, pauses, Beethoven's fifth, newspaper design, The Pompidou Center, artificial silence on telephones)

Yesterday and today [page 104]

There are loads of devices and services that keep track of the time and that alter their behavior when the next day begins. The decision to have this happen at midnight can confuse the users. Even more confusion is caused when the clocks go back an hour.

(Voice mail, baby-sitter, online TV guide, speaking clock, time differences, internet standard time, railways)

Documentaries [page 106]

We all know what TV documentaries are like, but what about interactive documentaries, is there such a thing? Could there be such a thing? If I present it then I am the one telling a story, but if you are interacting with it then are you the one telling the story?

(Muddy paths, the 'documentary engine')

PART FOUR: INTERACTION & SPECIFICATIONS

Labels [page 110]

People are always using labels, they help us when we are dealing with lots of things that are similar in appearance. What should we base our labels on when we create them? Are there useful labeling strategies in the real world and can they be used for the digital world?

(Moving house, file names)

Terminology [page 112]

The words that are used at the interface on buttons or in the help play a vital part in shaping the users experience and understanding of the system.

(MiniDisc recorders, Apple desktop interface, Prozac, Lymeswold cheese, hospital signs, fish)

International standards [page 114]

Different standard sizes in different countries cause great problems because they are usually the things that you least expect to be different.

(Filing cabinet folders, electrical plugs and sockets, Mars probe, spreadsheet programs)

Loops [page 116]

Information is handier if it is chunked up into discrete pieces. One way of arranging the collection is in a loop; a common abstract structure in computing information and in some real world things like carousel slide projectors.

(Parchment scrolls, piles of photographs, Jack Kerouac, fax machines, View Master, zoetrope, rolodex, Kodak carousel slide projector)

Sequences [page 118]

The linear sequence is one of the most common structures of information. What are the different abstract forms it can take? How can sequences be labeled and how does the user cope with labeling problems?

(Queues, numbering freeway junctions, numbering floors, Italian bakers)

People flow [page 120]

In designing public spaces the control of the flow of people is important. This is also true of the flow of visitors to a website. In both cases they have to be guided, stopped and encouraged.

(Church towers, the 'Porter' honey-bee gate, turnstiles, kissing gates, cattle grids, cat flaps, medicine bottles, speed bumps, kung-fu monastery)

Big choices [page 122]

As the range of consumer choice increases users are able to choose from ever larger collections of items. What are the problems associated with this task and how can they be overcome?

(Piggly-wiggly store, TV channels, single menu restaurants, storekeeper, reccomendations)

Broadcasting [page 124]

Broadcasting is sending a message from one source to many recipients. With digital technology that group of recipients can be very precisely chosen and profiled, how do different broadcast technologies define the target groups that they broadcast to?

(Bell-ringing, village bards, African drumming, smoke signals, spam email)

Information technology [page 126]

Organizing information is a discipline that has boomed with the advent of the computer, but when did we first start designing tools to help us with information? Was it as far back as the invention of printing?

(The hyperlink, book indexes, the Qur'an)

Specifications [page 128]

Specifications are the language that the designer uses to communicate to the builders/producers. Specifications for interactive systems are notoriously difficult. Are they necessary or are there advantages in working with informal briefings?

(Architecture, Lego blocks, Italian designers and model makers)

Shopping [page 130]

What are the building blocks of interaction when it comes to buying things, and what examples of these interactions are there in the real world?

(Children playing at shop, drink vending machine, restaurant, train tickets, stock options, Italian butchers)

Sound [page 132]

Different systems for dealing with sound have different structures and different ways that the user can manipulate and navigate those structures. From tapes to CD players some operations remain the same whilst new technologies introduce new functions like the CD 'shuffle' function.

(Cassette tapes, audio CDs, videotapes, DAT, MiniDisc, voice mail, digital wallet, Joni Mitchell)

Forms [page 134]

The design of forms whether on paper or online can have a great impact on the users experience and thus the effectiveness of the form itself.

(Tax/IRS forms, car smash)

Flexible systems [page 136]

As systems increase in size and complexity their ability to be flexible decreases. Also the move from paper to digital systems has the same effect. There are ways of maintaining this flexibility.

(Garden center delivery, credit card details)

APPENDIX C

The articles have been ordered according to their content, they have been brought together in sections that deal with specific areas of interaction design. For those readers who want to know about the order that they were actually published in, the details are in the table below.

Published title	First published	Where published	Title (in book)	page (in book)
Coffee	Jul-95	Bulletin Vol. 27(3)	Coffee	44
Buttons	Oct-95	Bulletin Vol. 27(4)	Buttons	32
Uppers and downers	Jan-96	Bulletin Vol. 28(1)	Elevators	48
Sticky labels	Apr-96	Bulletin Vol. 28(2)	Labels	110
Waiting	Jul-96	Bulletin Vol. 28(3)	Waiting	90
Cooking	Oct-96	Bulletin Vol. 28(4)	Food	56
Children	Jan-97	Bulletin Vol. 29(1)	Children	76
Powerful functions	Apr-97	Bulletin Vol. 29(2)	Powerful functions	40
Safety catches	Jul-97	Bulletin Vol. 29(3)	Safety catches	42
Cafés	Oct-97	Bulletin Vol. 29(4)	Cafés	54
Snoozing	Jan-98	Bulletin Vol. 30(1)	Snooze functions	88
Key states in ranges	Apr-98	Bulletin Vol. 30(2)	Slider controls	34
Protocols for commerce	Jul-98	Bulletin Vol. 30(3)	Shopping	130
Video conferencing	Oct-98	Bulletin Vol. 30(4)	Video conferencing	52
Video recorder	Jan-99	Bulletin Vol. 31(1)	Video	50
Pointing	Apr-99	Bulletin Vol. 31(2)	Pointing	78

Published title	First published	Where published	Title (in book)	page (in book)
The right length	Jul-99	Bulletin Vol. 31(3)	Length	98
Telephones	Oct-99	Bulletin Vol. 31(4)	Telephones	46
Paths	Jan-00	Bulletin Vol. 32(1)	Paths	96
Marks and scratches	Apr-00	Bulletin Vol. 32(2)	Marks and scratches	68
Junk	Jul-00	Bulletin Vol. 32(3)	Junk	66
Specifications	Sep-00	Bulletin Vol. 32(4)	Specifications	128
What's in a word?	Nov-00	Bulletin Vol. 32(5)	Terminology	112
Managing sound	Jan-01	Bulletin Vol. 33(1)	Sound	132
Nothing	Mar-01	Bulletin Vol. 33(2)	Blank	102
Contact	May-01	Bulletin Vol. 33(3)	Contact	62
International standards	Jul-01	Bulletin Vol. 33(4)	International standards	114
Plane living	Sep-01	Bulletin Vol. 33(5)	Stereo vision	72
Take-out service	Nov-01	Bulletin Vol. 33(6)	Take-out service	70
Loops	Jan-02	Bulletin Vol. 34(1)	Loops	116
Filters	Mar-02	Bulletin Vol. 34(2)	People flow	120
Two users – one preset	May-02	Bulletin Vol. 34(3)	A user group of two	64
Fill-in forms	Jul-02	Bulletin Vol. 34(4)	Forms	134
Modeling a process	Sep-02	Bulletin Vol. 34(5)	Flexible systems	136
Real time	Nov-02	Bulletin Vol. 34(6)	Real time	94
Interactive documentaries	Jan-03	Bulletin Vol. 35(1)	Documentaries	106
Is it on?	Mar-03	Bulletin Vol. 35(2)	Off and on	38
Goodbye	May-03	Bulletin Vol. 35(3)	Goodbye	100
Silly noises	May-03	Online	Funny noises	84
Broadcasting	Jul-03	Online	Broadcasting	124
Information technology	Sep-03	Online	Information technology	126
Infra red	Nov-03	Online	Infra-red	58
Sequences	Jan-04	Online	Sequences	118
Interruptions	Mar-04	Online	Interruptions	92
Left & right	May-04	Online	Left or right?	80
Volume control	Jul-04	Exclusive to book	Volume control	36
Being overheard	Jul-04	Exclusive to book	Being overheard	74
Do what I mean (not what I say)	Jul-04	Exclusive to book	Do what I mean (not what I say)	82
Yesterday & today	Jul-04	Exclusive to book	Yesterday & today	104
Big choices	Jul-04	Exclusive to book	Big choices	122

INDEX

F
facial expressions, 14
fax machines, 33, 63, 116
feedback, 44, 48, 57, 74, 90
files & directories, 66, 67, 110, 114
filing cabinets, 114
fire alarms, 92
fire extinguishers, 42, 43
flexible systems, 136
floppy disks, 93
fly-by-wire, 82
food, 56
forms, 121, 134
Freud, Sigmund, 14
fridges, 85
Furbies, 75

G
gaps, 129
gas burner, 34
graphic designers, 5
Guigiaro, Giorgetto, 129

H
hair-curler, 39
HAL, 43, 48
hanky-codes, 14
hardware, 31
HCI, 6
headphones, 71
hedgehogs, 121
hi-fi, 35
hospitals, 113
hot badges, 63
housing design, 4
human-human interaction, 13
   sources of, 16
hyperlinks, 126

I
indexes /indices, 127
industrial designers, 5
information, x
   loops, 116
   relevance of, 134
   structures, 117
information technology, 126
infra-red, 57
insurance forms, 135
intelligent agents, 123
intelligent systems, 46, 82, 83
interaction cycle, 22
interaction design, 5
   appreciation of, 6
   cost of, 139
   scope of, 6
interactions, 1
   context of, 22
   finishing, 100
   human-human, 13
   observing, 21
   quality, 18
interactive documentaries, 106
interactive games, 106
international standards, 114
internet cafés, 62
interruptions, 92
IP numbers, 32
iron, 39

J
joysticks, 35
junk, 66

K
Kerouac, Jack, 116
kettles, 38
key states, 34, 35

Other Lon Barfield titles published by Addison Wesley and Bosko Books

The User Interface: Concepts and Design

by Lon Barfield   *(Bosko Books 2004, ISBN: 0-9547239-0-2)*

Everybody has problems using technology, from heating controls through to video recorders. Move to computers and the problems are even worse; even the simplest computer programs seem to behave in strange ways. This book considers the problems of usability of technology and examines the factors that play a role in the design of such systems. Its goal is to introduce students and those working in related areas to the issues and to support them in analyzing problems and coming up with their own designs. It covers the issues surrounding the design of everyday technology before bringing computers into the picture and looking at how those issues change with the design of the user interface to computer systems. There are plenty of good seminar style exercises with accompanying guidelines.

The text uses numerous real-world examples to get its message across and it does so in an 'amusing and authoritative' style. It steers clear of technical issues which means that it is very general in nature, that it retains it's relevance as technologies change and that the text does not get bogged down in technical jargon. As well as the exercises, each chapter has an imaginary dialogue between Hemelsworth, a frustrated lord, and his dim-witted butler Barker, who is prone to behaving like your average computer system.

First printed and reprinted by Addison Wesley, this timeless title is now available from Bosko Books. It is still relevant and useful, and continues to be used to teach interaction design courses and computing courses relating to the user interface.

More information and extracts at: *www.idhub.com/concepts*

Design for New Media; Interaction Design for Multimedia and the Web

by Lon Barfield   *(Addison Wesley 2003, ISBN: 0-201-59609-1)*

Where do you start when designing interactive digital media such as a website? What are the building blocks, the key elements? How do you bring them together? How do you start your design?

This book answers these questions, and many more besides. In simple, non-technical terms it outlines the basic elements of design for new media: text, layout, icons, sound, color, video and animation. As well as considering each of these elements in the context of interaction design the book goes on to show how they are combined into an interactive whole using narrative, structure, navigation, contexts. Finally, the process of actually designing interactivity is dealt with; from generating ideas through to specifications and prototyping.

A key textbook for students designing interactive, digital media, and for practitioners who want a grounding that is solid and smooth to read. This book presents a well-structured and uniquely entertaining guide to what today's web or multimedia designer needs to know about.

More information and extracts at: *www.idhub.com/newmedia*

Forthcoming titles from Bosko Books [ISBNs and titles subject to change]

Digital Media for Art & Design

by Lon Barfield   *(Bosko Books 2004, ISBN: 0-9547239-3-7)*

Digital Media for Art & Design introduces students to the concepts, technologies and tools involved in the creation of digital media.

The number of courses connected with digital media is on the increase. Even conventional courses such as graphic design now incorporate a significant amount of digital work. Practitioners too are being called upon to get involved in an ever-increasing variety of digitally-based tasks.

As computers get more powerful, the designers and developers need to control this great power that is being unleashed and to turn it to their own ends. What sort of books can help them do this? Huge technical texts can become part of the problem. Glossy, coffee-table books show what can be done, but don't help you to do it. Detailed how-to-do-it books take you through step-by-step recipes that you can trot out like a dog doing tricks, but they don't give you what you need to be able to do it yourself.

What is needed is a text that covers the area, but does so in an informed and non-technical manner. A text that gives the readers a handle on the many possibilities that digital media has to offer, and explains those ideas that are necessary. That is the goal that this book fulfills; to give the student an overview, and to give them conceptual tools which allow them to make informed decisions. From image formats to key Photoshop concepts, from file sizes for digital audio to scripting for websites, from bitmaps and vectors to programming in Flash. A range of key topics are covered on a 'need to know' basis; technical details are only included if they are relevant and are explained in simple and direct terms. This books is a valuable resource for anyone getting to grips with the world of digital media.

More information at: *www.idhub.com/digitalmedia*

Got something inspirational to say about design?

Then maybe you should say it. Bosko Books is constantly on the look out for experts to contribute to current book projects or to put together books of their own. We publish books and digital products to support designers and those that teach design. We aim to serve markets related to interaction design, covering areas that include:

- Web design
- Information architecture
- Industrial design
- Digital media
- Architecture and town planning
- Digital technologies
- Information technology

So if you have an idea for a book, or there is an area that you feel is relevant and about which you want to write, then get in touch:

info@boskobooks.com

Or go to:

www.boskobooks.com/write

www.ingramcontent.com/pod-product-compliance
Lightning Source LLC
Chambersburg PA
CBHW051411200326
41520CB00023B/7190